A MANUAL OF HUMAN ANATOMY

VOLUME III

THE UPPER AND LOWER LIMBS

A MANUAL OF HUMAN ANATOMY

VOLUME III

THE UPPER AND LOWER LIMBS

By

J. T. AITKEN, M.D.
PROFESSOR OF ANATOMY AT
UNIVERSITY COLLEGE, LONDON

G. CAUSEY, M.B., F.R.C.S.
PROFESSOR EMERITUS,
UNIVERSITY OF LONDON

J. JOSEPH, M.D., D.Sc., F.R.C.O.G.
PROFESSOR OF ANATOMY AT
GUY'S HOSPITAL MEDICAL SCHOOL,
LONDON

J. Z. YOUNG, M.A., D.Sc., LL.D., F.R.S.
PROFESSOR EMERITUS
UNIVERSITY OF LONDON

THIRD EDITION

CHURCHILL LIVINGSTONE
EDINBURGH LONDON AND NEW YORK
1976

CHURCHILL LIVINGSTONE

Medical Division of Longman Group Limited

Distributed in the United States of America by
Longman Inc., 72 Fifth Avenue, New York,
N.Y. 10011, and by associated companies,
branches and representatives throughout
the world.

First Edition . .	1957
Second Edition . .	1964
Reprinted . .	1968
Reprinted . .	1972
Third Edition . .	1976

ISBN 0 443 01242 3

Library of Congress Cataloging in Publication Data

Aitken, John Thomas.
A manual of human anatomy.

Includes indexes.

CONTENTS: v. 1. Thorax, abdomen, and pelvis.—v. 2. Head and neck.—v. 3. The upper and lower limb.
1. Anatomy, Human. I. Title. [DNLM: 1. Anatomy, Regional. 2. Dissection. QS130 A311m]
QM23.2.A35 1975 611 74-33179

PRINTED IN GREAT BRITAIN

PREFACE TO THE THIRD EDITION

SINCE the publication of the Second Edition in 1964, there have been two reprints of the volumes. The continued confidence shown by this demand is most gratifying to the authors and justifies their original conception of the scope of a preclinical course in topographical anatomy.

In many departments radical changes are being made in the methods of teaching anatomy and in the content of the course. These experiments are to be welcomed. Usually the time available for careful dissection has been reduced and many students have to be content with a more rapid and less detailed approach to the body. The authors suggest that the dissection of some regions may be omitted, although the dissecting instructions are retained in the text.

In this edition, the number of volumes has been reduced and the contents re-arranged. Volume I contains the thorax, abdomen and pelvis, Volume II the head and neck and Volume III the limbs. It is hoped that this arrangement will keep the total cost as low as possible and not inconvenience students or departments.

We are grateful to colleagues and students for helpful suggestions, and to the staff of Churchill Livingstone who continue to be most co-operative.

THE AUTHORS.

LONDON
1975

PREFACE TO THE FIRST EDITION

THE purpose of these Manuals is to give the student of human anatomy a method of dissecting the body and to act as a guide to the extent of the knowledge expected of him in the second medical examination. An attempt has been made to link together the structure and function of the different parts of the body, and the anatomy necessary for a future study of clinical medicine or an understanding of the development of the part is emphasised. Paragraphs indicating the functions of the parts under consideration have, where appropriate, been introduced after the practical instructions and topographical details. The study of each part can thus be undertaken with some knowledge of the functional implications of the anatomy and not as a mere exercise of memory. Much detail has been omitted.

A co-ordinated course is more easily organised if all the members of the class are dissecting the same part at the same time and the instructions in the Manuals are presented on this assumption. A most important part of the teaching is carried out by means of small classes on osteology and surface anatomy. For these classes we have found it useful to indicate what the students should know. The students do the work themselves and it is then checked by a demonstrator. Appropriate lists for such work are found at the end of each volume.

It has been found advantageous to begin with the dissection of the thorax. This results in an early acquaintance with the heart and lungs and with the peripheral and autonomic nervous systems, all of which receive attention in most introductory courses of physiology. From the thorax, the student proceeds to dissect the upper limb (vol. I), the head and neck (vol. II), the abdomen and pelvis (vol. III) and the lower limb (vol. IV). The descriptions and instructions in the Manuals assume that this order has been followed. Instructions for the use of the Manuals where a different order is employed are given opposite page 1. The gross and histological structure of the brain and spinal cord are described in vol. V.

PREFACE TO THE FIRST EDITION

Each part of the body is subdivided for convenience into smaller regions. In the limbs these regions centre around the joints and in other parts around the larger morphological or functional units. In each region, a short introduction is followed by dissecting instructions, including a description of many of the structures being dissected. There follow paragraphs on further details and relations of the structures, and their functions.

Summaries of the cutaneous nerve supply and of the lymphatic drainage of the part dissected are found towards the end of each section of the Manual.

In the early stages of the planning and writing of these Manuals, Dr. W. A. Fell, now of Addenbrooke's Hospital, Cambridge, and Dr. D. H. L. Evans of University College, London, contributed to the work and much helpful criticism has been received from other colleagues and students.

The illustrations were produced by Miss E. R. Turlington and Miss J. de Vere, largely from specimens and drawings in the Anatomy Department at University College, London. As the main object of the pictures is to illustrate the text, all unnecessary complicating details have been omitted and the salient features emphasised by the use of colour.

Our thanks are also due to Miss A. Baxter and Miss M. Lynn for typing the final draft of the Manuals, and the staff of E. & S. Livingstone for the production and publication of the Manuals.

THE AUTHORS.

LONDON,
FEBRUARY, 1957

CONTENTS

NOTE

THE order in which the different parts of the body are dissected varies. Many departments prefer all their students to dissect the same part at the same time. By beginning with the thorax (Vol. I) the students are quickly introduced to the organs of respiration and circulation, and also the spinal and autonomic parts of the nervous system. From the thorax, dissection can proceed to the abdomen and pelvis, and later to the neck and head (Vol. II), and then into the limbs.

If the body is dissected in the sequence of these Manuals some instructions may appear to be superfluous. However, in some departments, where different groups of students dissect the various parts of the body at the same time, some rearrangement of the order is required. If dissection is begun with the body on its back, then dissectors of the upper limb (Volume III, Chapter 3) should collaborate with dissectors of the thorax (Volume I, Chapter 2). Those dissecting the lower limb (Volume III, Chapter 16) should collaborate with dissectors of the anterior abdominal wall (Volume I, Chapter 10).

If dissection is begun with the body on its face then again those dissecting upper limb (Volume III, Chapter 5) collaborate with those dissecting head and neck (Volume II, Chapter 2). Those dissecting the lower limb (Volume III, Chapter 15) collaborate with dissectors of the head and neck (Volume II, Chapter 2).

ORIENTATION

TO help in the description of a structure or a region certain terms are used and they have an agreed interpretation. The **anatomical position** is one in which the person stands upright, with the feet together, the eyes looking forward, and the arms straight along the sides of the body and the palms of the hands directed forwards. The front of the body is called the **anterior** surface and the back is called the **posterior** surface (see outside front cover drawing). The terms **ventral** and **dorsal** may be used for the front and back respectively. Higher structures are **superior** and lower structures are **inferior. Median** structures are found in the midline of the body and the terms **medial** (nearer to) and **lateral** (further from) are relative to the midline.

A **sagittal plane** passes vertically anteroposteriorly through the body and movements in this plane (see inside front cover drawing) are called **flexion** (forwards) or **extension** (backwards). During development, the lower limb distal to the hip joint becomes rotated so that forward movement at the knee, ankle and foot joints is called extension and backwards movement is called flexion. A vertical plane at right angles to the sagittal is called a **coronal (frontal) plane.** Movements of the limbs in this plane are called **adduction** (towards the midline) and **abduction** (away from the midline). At certain joints, **rotation** occurs about a longitudinal axis.

THE UPPER AND LOWER LIMBS

CHAPTER 1

GENERAL INTRODUCTION

THE limbs are basically concerned with movement and posture. The movement may be mainly prehension (grasping and exploring) as in the upper limb of man or locomotion as in the lower limb. Though in quadrupeds all the limbs are used almost exclusively for locomotion, in bipeds other functions appear in the non-weightbearing limbs. In some animals (*e.g.* brachiators), the upper limbs are specialized for locomotion as well as prehension. Limbs have also important sensory functions. This is obvious in the hand, but the foot also can detect irregularities and changes in the inclination of the surface of the ground.

The anatomical pattern of the upper and lower limbs is similar. The pectoral girdle (scapulae and clavicles) connects the upper limbs to the upper part of the trunk, and the pelvic girdle (hip bones) connects the lower limbs to the lower part. The proximal segment of both limbs has one long bone (humerus or femur) and the intermediate segment has two bones (radius and ulna or tibia and fibula). The distal segment (hand or foot) consists of the carpus or tarsus (wrist or ankle regions), the metacarpals or the metatarsals, and the phalanges (the bones of the digits).

The upper limb is characterized by its mobility and its sensitivity. Most of its joints are very freely moveable and the hand is especially important in exploring the local environment and in holding tools. Almost all parts of the surface of the body are ordinarily accessible to one or other hand. The lower limb is mainly involved in locomotion and posture. Its joints are more stable and some have a more limited range of movements. The pectoral girdle is much less rigidly attached to the trunk and contributes considerably to the mobility of the upper limb.

Artificial limbs are made to carry out the basic functions of the missing part. An artificial hand which may be aesthetically acceptable may have little functional value. An artificial lower limb, however, can carry weight and allow movement, and also be aesthetically acceptable.

A limb first appears on the body wall of the embryo as a fold of skin into which mesoderm grows so that a limb bud is formed. This occurs during the 4th-5th weeks of embryonic life. The upper limb fold of skin is opposite the 4th cervical to the 2nd thoracic segments of the body and the lower limb fold is opposite the 2nd lumbar to the 3rd sacral segments. In man the continuity of the limb muscles with the individual trunk myotomes is not clear, nevertheless evidence of the segmental origin is seen in the distribution of the nerves to the skin (Figs. 23 and 62).

In the early stages, each limb bud has a ventral and dorsal surface. The 1st digit (thumb or big toe) is on the headward side of the limb. The limbs grow outwards in the coronal plane of the trunk and then are assumed to move anteriorly towards the midline in the horizontal plane.

The upper limbs rotate outwards (laterally) and then backwards so that the palmar surface comes to face forwards (anteriorly). The lower limbs rotate inwards (medially) and then backwards so that the sole of the foot faces backwards (posteriorly). As a result of the rotation in the upper limbs the thumb is on the lateral (radial) side of the hand, and in the lower limb the big toe is on the medial (tibial) side of the foot. If the forearm is rotated so that the palm faces backwards and is placed over the upper surface of the foot on the same side, then the general arrangement of the structures is similar in the upper and lower limbs. The rotation of the lower limb is regarded as occurring in the thigh distal to the hip region.

Ossification of a long bone of a limb begins in a primary centre in its body (shaft) which appears about the eighth week of intra-uterine life. Secondary centres of ossification in the ends of the bone mostly appear after birth. Growth in length continues until about the eighteenth year in men, but finishes two years earlier in women.

Movements at joints

The following terms are commonly adopted for the description of movements at joints from the anatomical position (see inside front cover).

(1) Angular movements:

(*a*) abduction and adduction, away from or towards the midline, the movements occurring in a coronal plane round an anteroposterior axis.

(*b*) flexion and extension, forwards and backwards in a sagittal plane, the movements occurring round a transverse axis at right angles to the axis in (*a*).

(2) Circumduction, in which the distal end of one bone describes the base of a cone whose apex is at the proximal end.

(3) Rotation, in which a bone is rotated about its own long axis.

(4) Gliding, in which there is no rotatory or angular component.

Other descriptive terms will be defined in later sections of the book.

Plan of dissection

The dissection and description of the limbs will be organised round the joints. As far as possible, dissections of the two limbs will keep "in step", so that if groups of students are dissecting upper and lower limbs simultaneously then reference can be made from one to the other. Examination of the shoulder girdle may take some extra time, though this region may have been started when the thorax was dissected. At the end of Part I, Upper Limb (page 90) and of Part II, Lower Limb (page 198) are outlines for classes in osteology and surface (living) anatomy.

PART I - UPPER LIMB

CHAPTER 2

FUNCTIONS OF THE UPPER LIMB IN MAN

THE upper limbs are based on a plan of structure arranged originally for purposes of locomotion but they have now almost wholly lost that function and become instead organs for sensation and grasping and holding, so adjusting the environment to suit man's needs. The movements depend for their efficacy as much on their delicacy as on their force. An appreciation of the extent to which this is true in people with different occupations is of first importance in assessing the way in which they are likely to want to use a damaged limb and hence in deciding on the treatment which may be best employed. On the sensory side it is important to know the innervation of the various parts of the skin so as to understand the degrees of impairment of sensory function likely to result from injury to each nerve. Finally, the texture and properties of the skin of the hands are of special importance.

The use of the hands for delicate and accurate movements is made possible by three factors, (1) the great freedom of movement of the limb in all directions, (2) the large part of the cerebral cortex that is concerned with movements of the hands and fingers, (3) an adequate sensory supply to give information about the objects which are being handled.

The use of the limb for applying force for the movement of objects depends on rather different factors, (1) the ability to hold joints at angles that allow the muscles to work at a proper mechanical advantage, (2) the strength of these muscles, of the tendons by which they are attached, and of the bones, (3) the provision of a stable base in the trunk and lower limbs.

This brief survey of the functions of the upper limb will emphasize the points to which special attention must be directed. Apart from the cerebral control of the upper limb, the most outstanding of these is evidently the mobility of the joints. The angles through which the parts can move, the importance for particular activities of each type of mobility and the dangers of dislocation and limitation of movement present at each joint must be studied. The muscles which move these joints must also be examined and the nature of the movements associated with these muscles determined. Though most of the limb muscles cross only one joint, a considerable number cross two joints and so have more complicated actions. The bones act not only as levers for the muscles but also as rigid objects upon which the main strain is exerted if the limb is subjected to unusual external forces in falling or other accidents.

In the anatomical position the upper limbs are held at the sides, with the palms directed forwards. The lateral border of the limb represents the **pre-axial** and the medial border the **postaxial** aspects of the limb. The value of understanding this is that the headward nerve roots of the brachial plexus supply the skin of the pre-axial aspect. Thus the 4th to the 6th cervical spinal nerves supply the lateral (pre-axial) aspects, and the 8th cervical to the 2nd thoracic nerves supply the medial (postaxial) aspects. The 7th cervical nerve is the central nerve (Fig. 23) and supplies the skin of the middle finger.

The upper limb is attached to the axial skeleton by the pectoral girdle and consists of three segments, (1) the upper arm, (2) the forearm, (3) the hand. The **axilla** (armpit) is the hollow between the chest wall and the arm. The **cubital fossa** is the hollow of the elbow. The forearm is said to have anterior (palmar) and posterior (dorsal) surfaces and lateral (radial) and medial (ulnar) borders. The hand has palmar and dorsal surfaces. The ball of the thumb is the **thenar eminence** and on the ulnar border of the palm of the hand is the **hypothenar eminence.** The digits are: I. pollex (the thumb), II. index (the forefinger), III. medius (the middle finger), IV. anularis (the ring finger), V. minimus (the little finger). There are two segments (proximal and distal phalanges) in the thumb and three (proximal, middle and distal phalanges) in each of the other fingers.

THE MOVEMENTS AT THE JOINTS

The functions of the upper limb as a sense organ and as a manipulating instrument require above all else a high degree of mobility, and several of the joints are not limited to simple movements in one plane but allow movements in several directions. The descriptive terms are illustrated on the outside and inside of the front cover.

The shoulder joint with the shoulder (pectoral) girdle allows a wider range of movement than any other joint in the body. Starting from the anatomical position flexion is forward movement through about 180°. The backward movement (extension) is limited to about 30°. Abduction has a range of about 180°, producing almost the same final position as full flexion. Adduction when not limited by contact with the side of the body has a range of about 30°. Movements in the various directions are combined to produce circumduction. Medial and lateral rotation of the limb at the shoulder joint together amount to about 180°.

The elbow joint has, by comparison, a limited range of movement. From the anatomical position a considerable degree of flexion is possible, but little or no extension. The joint, in association with the joints between the radius and the ulna, also allows rotation of the forearm (see inside front cover). In the anatomical position, with the thumb laterally, the forearm and hand are said to be supinated. The palmar surface can be turned through 180° in a medial direction (pronated) so that the palm is directed backwards and the thumb medially. Rotation of the hand through more than 180° involves rotation of the humerus at the shoulder.

The wrist joint allows flexion and extension, abduction and adduction (also known as radial and ulnar deviation respectively), and circumduction. Movement at the joints between the palm and the fingers includes flexion, a little extension, adduction, abduction (by which the various fingers are spread away from each other) and circumduction. In describing some of these movements, the middle finger is taken as an axis, movement of a digit away from it being called abduction and towards it, adduction. The interphalangeal joints allow only flexion from the anatomical position.

The independence of movement of the thumb arises from the fact that it is separated from the palm and that the joint at its base is saddle-shaped and allows flexion, extension, abduction, adduction and rotation. The tip of the thumb can be brought into contact with the other fingers across the palm, a movement known as opposition.

By means of the movements at the various joints the arm is able to fulfil its great variety of functions. It can seize heavy objects and move them in various directions (though not of course without movements of the body) and also hold small objects and manipulate them with accuracy, as in writing. It can reach most parts of the body or in any direction to feel objects nearby. Contrast this freedom of movement with the limitations in the lower limb and in a mammal such as a horse that uses its forelimbs mainly for locomotion and support.

CHAPTER 3

THE ANTERIOR ASPECT OF THE SHOULDER GIRDLE

INTRODUCTION

THE great range of movement at the shoulder is produced in two ways. Firstly there is a nearly flat saucer-like facet on the scapula for the spherical head of the humerus, so that the latter can move very freely. Secondly this range of movement is increased by moving the whole shoulder girdle so as to change the direction of the facet for the humerus. In considering the shoulder joint, therefore, we have to think also of the scapula and the muscles that move it, as well as those which move the humerus on the scapula. A system designed to allow such great freedom of movement is especially likely to suffer from (1) conditions which cause limitation of this movement and (2) dislocations on account of weaknesses inherent in an arrangement able to give such freedom. Actually in spite of its freedom the arm is very firmly anchored to the body by a number of muscles. Some of these lead directly from the axial skeleton (vertebrae and ribs) to the humerus. Others fix the humerus to the scapula, which is in turn held by muscles to the axial skeleton.

The muscles to be examined in this dissection pull the limb across the front of the chest and also are involved when the limb is raised above the head. The hollow between the chest wall and the limb (the axilla) contains the principal vessels and nerves supplying the limb and has to be carefully dissected. Also in the axilla will be found lymph nodes which drain the limb and the chest wall. The grouping of these nodes is more apparent in pathological states or in specially prepared material than in most dissecting room bodies.

On the cadaver and the skeleton identify the following bony features: the clavicle, the sternum, the jugular notch (at the upper end of the sternum), the xiphoid process (at the lower end of the sternum), the costal (rib) margin, the ribs and costal cartilages. Between the ribs are the intercostal spaces. The **clavicle** can be felt along its whole length and has two curves, one forwards (medially) and one backwards (laterally). The site of the **acromioclavicular joint** can be felt laterally. The clavicle is more prominent on the upper aspect of the joint than is the acromion. In the hollow below the lateral third of the clavicle, palpate the **coracoid process** of the **scapula.** The upper surface of the **acromion** of the scapula is felt lateral to the acromioclavicular joint and still more lateral is the projection of the upper end of the **humerus,** the **greater tuberosity,** which with the overlying muscle (deltoid) produces the characteristic rounded contour of the shoulder. Note also the **head** of the humerus and the **lesser tuberosity.** The head of the humerus articulates with the **glenoid cavity** of the scapula and is joined to the shaft by the **anatomical neck.** The region of the shaft below the tuberosities is called the **surgical neck** because of the frequency of fractures here. Between the two tuberosities is the **intertubercular groove.** Hold the upper limb away from the body and grasp between the thumb and forefinger the anterior wall of the axilla containing the pectoralis major muscle. Similarly grasp the posterior wall of the axilla containing the teres major, latissimus dorsi and subscapularis. Run two fingers along the medial wall of the axilla, formed by the ribs covered by the serratus anterior, and along the lateral wall in which the body of the humerus and the main vessels and nerves of the arm can be felt. The apex of the axilla is bounded by the clavicle in front, the scapula behind and the first rib medially. This aperture is the means whereby the structures pass between the neck and the upper limb.

DISSECTION

Abduct the limb to 90° and fasten it on to a board lying under the shoulders. If the thorax has not been dissected, make the following incisions through the skin (see back cover):

1. from the jugular notch to the lower end of the sternum,
2. from the jugular notch along the length of the clavicle to the tip of the shoulder.
3. from the lower end of the sternum along the costal margin to the midaxillary line.

If the thorax has been dissected, turn to page 13.

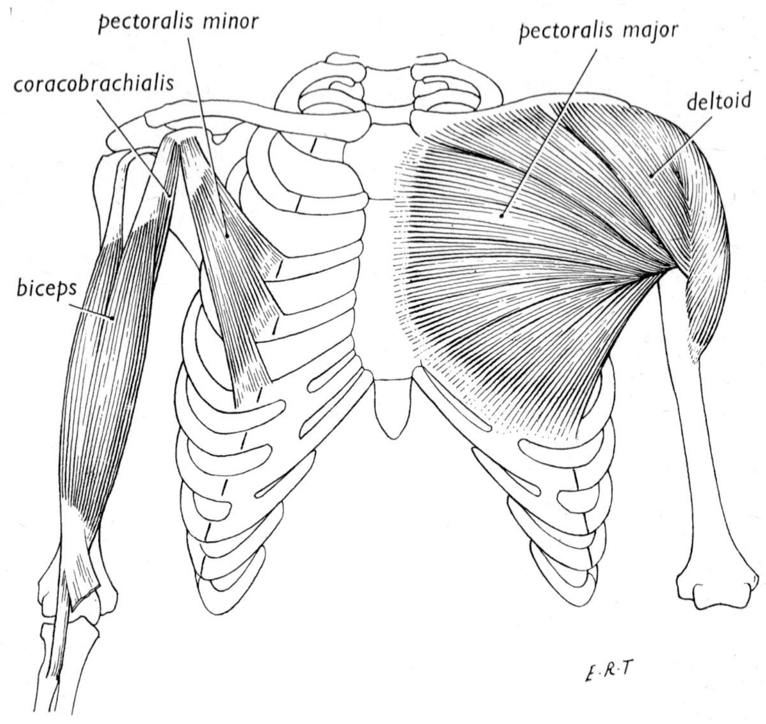

FIG. 1
On the right side, the pectoralis major and deltoid muscles have been removed to expose the attachments of pectoralis minor, coracobrachialis and biceps brachii.

The superficial structures

With forceps take hold of one corner of the flap of skin thus marked out and dissect it away from the underlying tissue. In this manner the flap should be reflected laterally and backwards as far as the point of the shoulder and as far as the lowest point of the costal margin. The pigmented part of the skin around the nipple, the **areola,** is much thinner. After removing the skin, the underlying subcutaneous tissue, called the **superficial fascia,** is exposed; it often contains much fat. Though it appears in the cadaver as a firm and tough layer, in the living body the fascia is loose and not firmly attached to the skin, and at body temperature the fat globules are fluid. The superficial fascia thus allows movement of the skin on the underlying structures. In certain situations such as the palm of the hand and the sole of the foot where such movement would be disadvantageous, the skin is firmly attached to the deeper tissues.

In the superficial fascia inferior to the clavicle some cutaneous nerves, the supraclavicular branches of the 3rd and 4th cervical nerves, may be found along with some muscle fibres spreading on to the chest from the platysma muscle of the neck. The rest of the skin of the front of the chest wall is supplied by the thoracic nerves, through the intercostal branches.

The **mammary gland** lies in the superficial fascia. In the male it remains rudimentary throughout life. In the female it changes considerably with age and increases in size during pregnancy and lactation. It is commonly the site of disease and therefore although it cannot be properly dissected even in female cadavers, its anatomy must be known in some detail (Vol. I, page 17).

With a scalpel and forceps remove the exposed superficial fascia. Small blood vessels and nerves are seen entering its deep surface on their way to the skin. The underlying **deep fascia** is now exposed. This forms an investing layer for practically the whole of the body. In the dissected specimen it appears white, fibrous and tough, owing to the action of the preserving fluid, but in the living body it is almost transparent, slippery and smooth. This again allows some movement of one structure on another. The deep fascia forms the outer connective tissue covering (epimysium) of the underlying muscle, the **pectoralis major** (Fig. 1). Examine the attachment of one head of the muscle to the medial half of the front of the clavicle and of the other head to the front of the sternum, to the upper six costal cartilages and to the aponeurosis on the front of the anterior abdominal wall. The other end of the muscle is attached to the upper end of the humerus and some difficulty may be experienced in exposing the tendon at this stage of the dissection.

Detach the pectoralis major from the clavicle, sternum and costal cartilages and turn the whole muscle laterally. While this is being done, pectoral nerves and vessels supplying the muscle should be cut as they enter its deep surface. By turning aside the pectoralis major the superficial layer of the anterior wall of the axilla is removed and the deep layer exposed. This consists of the clavicle above, the pectoralis minor below, and a sheet of fascia passing between these two structures, the **clavipectoral fascia.** Inferior to the clavicle and passing from it medially to the 1st rib is the subclavius muscle, which is enclosed by the clavipectoral fascia. The lateral pectoral nerve, the cephalic vein and branches of the axillary artery can be seen piercing the upper part of the clavipectoral fascia. **Pectoralis minor** arises from the 3rd, 4th and 5th ribs and passes upwards and laterally to the coracoid process of the scapula. Cut through its costal attachment and turn it laterally.

After reflecting pectoralis minor, identify three muscles that are attached to the ribs. The first, at the side of the midline, is the vertical **rectus abdominis,** whose upper end arises from the xiphoid process and the adjacent costal cartilages. The second is the horizontal **serratus anterior,** arising on the lateral side of the chest from the upper eight ribs. The third is the **external oblique** of the abdominal wall which is attached laterally to the lower eight ribs and interdigitates with the origin of the serratus anterior.

The dissection should continue by exposing the humeral attachment of pectoralis major. Incise the skin along the middle of the front of the upper arm from the point of the shoulder to the elbow, and reflect it downwards from the axilla and from the medial and superior parts of the upper arm. Remove any superficial fascia on the front of the chest wall in order to show that the deep fascia is continuous with the deep fascia in the axilla. Remove this fascia and expose the lateral part of the pectoral major, together with the front part of the deltoid muscle (Fig. 1). Between the two is a space called the **infraclavicular fossa,** in which is found the cephalic vein passing upwards to the axillary vein. The cephalic vein is the lateral of the two large veins draining the superficial tissues of the upper limb (Fig. 12). In the fossa, acromiothoracic branches of the axillary artery are also found. Trace the tendon of the pectoralis major to its attachment on the humerus. The **pectoralis minor** should be examined. It has costal attachments to the 3rd, 4th and 5th ribs, and its attachment to the coracoid process should now be cleaned. The muscle pulls the scapula forwards on the chest wall and rotates it about an anteroposterior axis, so that the inferior angle moves medially and the glenoid cavity is directed downwards. It may act as an accessory muscle of respiration if its scapular end is fixed. The medial and lateral pectoral nerves, branches of the medial and lateral cords of the brachial plexus respectively, give branches to the pectoral muscles.

Unless the middle third of the clavicle is removed the whole of the axilla cannot be fully exposed and the continuity of the structures in the axilla with those in the root of the neck cannot be seen. Until the neck is dissected it is best, however, to leave the clavicle intact or make only a single saw-cut at the level of the infraclavicular fossa. By abducting the arm as much as possible a fairly good view of the axilla can be obtained. The axilla contains a considerable amount of fat, which must be removed carefully. Some lymph nodes embedded in the fat are usually found and these should also be removed.

On the chest wall the attachment of the **serratus anterior** by means of a series of muscular slips to the upper eight ribs has been identified. Look for the nerve to the serratus anterior (long thoracic nerve) passing down on its superficial surface about midway between the anterior and posterior axillary folds. This nerve comes from the roots of the brachial plexus (5th, 6th and 7th cervical nerves) which lie in the neck. The **intercostobrachial nerve** in the axilla emerges from the second intercostal space, is a branch of the second intercostal nerve, and passes to the upper, medial aspect of the arm.

The lateral wall of the axilla should now be cleaned. Trace the **coracobrachialis** and **short head of biceps brachii** from the upper arm to the coracoid process. The coracobrachialis lies more medially and is attached to the medial side of the body of the humerus. Its main action is to flex the humerus. A large branch of the brachial plexus, the **musculocutaneous nerve,** passes through it, supplying the muscles of the front of the upper arm and the skin of the lateral side of the forearm.

STRUCTURAL DETAILS

The clavicle

This bone and the scapula form the shoulder girdle and are the main means whereby the upper limb is attached to the trunk. The girdle also moves on the trunk and thus increases the mobility of the upper limb. The clavicle articulates medially with the sternum and

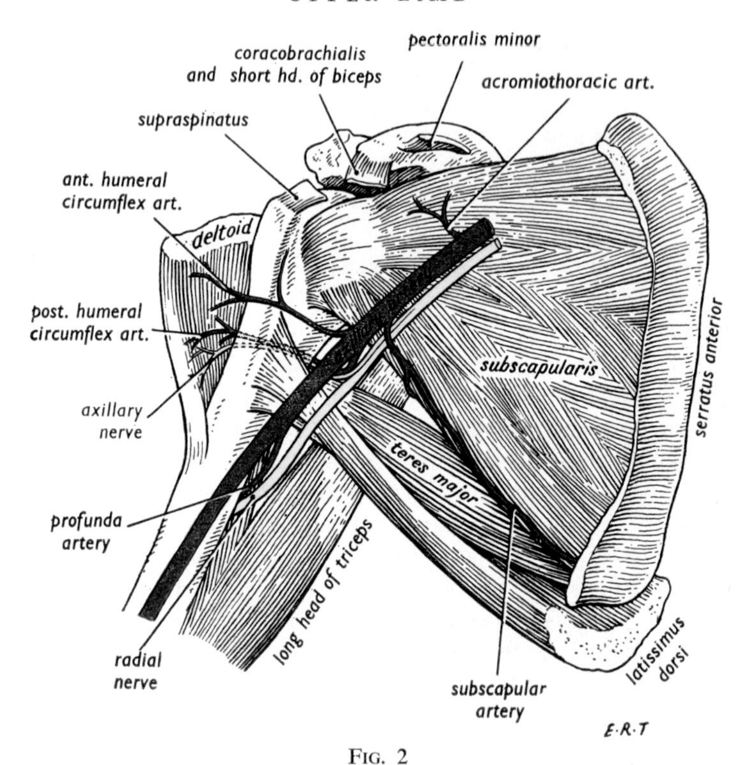

FIG. 2
Structures forming the posterior wall of the axilla. The artery and the humerus have been displaced laterally to expose the posterior cord of the brachial plexus.

1st costal cartilage and laterally with the acromion of the scapula. The sternal end is rounded and the acromial end is flat in section. The lateral third of the bone is convex backwards and the medial two thirds convex forwards. By examining the curves it is possible to determine which is the anterior and which is the posterior surface of the bone. The inferior surface shows a roughening at each end and a groove in the long axis of the bone, situated about its middle. Thus the upper surface can be distinguished from the lower. It is now possible to determine to which side of the body a clavicle belongs.

The clavicle **ossifies** in mesoderm from a primary centre in the body, appearing early in the 2nd month of intra-uterine life.

It is the first bone to ossify. A secondary centre in its sternal end appears about the 18th year and joins the body at about the 25th year (Fig. 10).

The clavicle acts as a strut holding the scapula and limb away from the trunk, thus assisting the free movement of the limb in all directions. The clavicle is better developed in man than in quadrupeds. Forces act mainly along the length of the human clavicle. Unlike the long bones of the limbs, it has no medullary cavity.

The pectoralis major

The attachment of the pectoralis major to the clavicle, sternum and costal cartilages should be studied again. The two portions, **clavicular** and **sternal,** pass laterally under the anterior edge of the deltoid and they are attached to the lateral lip of the intertubercular groove of the humerus. Both heads pull the humerus towards the trunk in the coronal plane (adduction) and rotate it medially about a longitudinal axis (medial rotation). The clavicular head can raise the humerus forwards in the sagittal plane (flexion) and the sternal head can pull the humerus downwards in the same plane (extension). With the humerus fixed the pectoralis major pulls the trunk up over the arms, as in climbing a rope. It may act as an accessory muscle of respiration if the humerus is fixed. Its nerve supply is from the medial and lateral pectoral nerves.

The serratus anterior

The serratus anterior passes backwards round the chest wall from the upper eight ribs to the medial border of the scapula (Figs. 2, 9 and 11). The lower four digitations pass to the inferior angle. The action of this muscle will be fully discussed with the movements at the shoulder joint. It can pull the whole of the scapula forwards round the chest wall, thus lengthening the reach of the upper limb, and it rotates the scapula by pulling on the inferior angle so that the glenoid cavity is directed upwards. Since the inferior angle moves laterally, this movement is called lateral rotation. This muscle may also be an accessory muscle of respiration if the scapula is fixed.

CHAPTER 4

THE CONTENTS OF THE AXILLA

INTRODUCTION

WITHIN the confines of the axilla are many important structures. The nerves and arteries can be dissected but the lymph nodes are not easily observed in dissecting room material. The nerves and arteries enter the axilla from the neck and thorax above the first rib (Fig. 3). They are embedded in fascia, continuous with that in the neck and forming a sheath for the structures. The axillary artery, a continuation of the subclavian artery, may be regarded as the central structure with the other vessels and nerves arranged round it. The brachial plexus is the means whereby the nerve fibres in the spinal nerves are distributed through the branches of the plexus to the limb. The basic pattern of the plexus can be seen in most bodies but there is considerable variation in the detailed arrangement. The plexus is usually derived from the 5th cervical to the 1st thoracic nerves of the spinal cord, but may be **prefixed** (4th cervical to 8th cervical nerves), or **postfixed** (6th cervical to 2nd thoracic nerves). In general the upper spinal nerves supply muscles in the more proximal parts, *e.g.* the shoulder girdle, and the lower nerves, muscles in the more distal parts, *e.g.* the hand.

The axillary artery gives off several large named branches that have a fairly constant course and distribution but there are also numerous small variable branches supplying the walls of the axilla and the lymph nodes. Some large branches pass to the scapular region where they anastomose with large branches of the subclavian artery. These anastomoses may form a collateral blood supply to the upper limb if the proximal part of the axillary artery is obstructed by disease or at operation.

Define the walls of the axilla in the cadaver and living subject. In the anterior wall are the pectoralis major superficially and deep to it the pectoralis minor (page 11). The medial wall is formed by the serratus anterior and the upper ribs. Laterally are the humerus, the coracobrachialis and the short head of the biceps; posteriorly are three muscles attached to the upper end of the humerus,

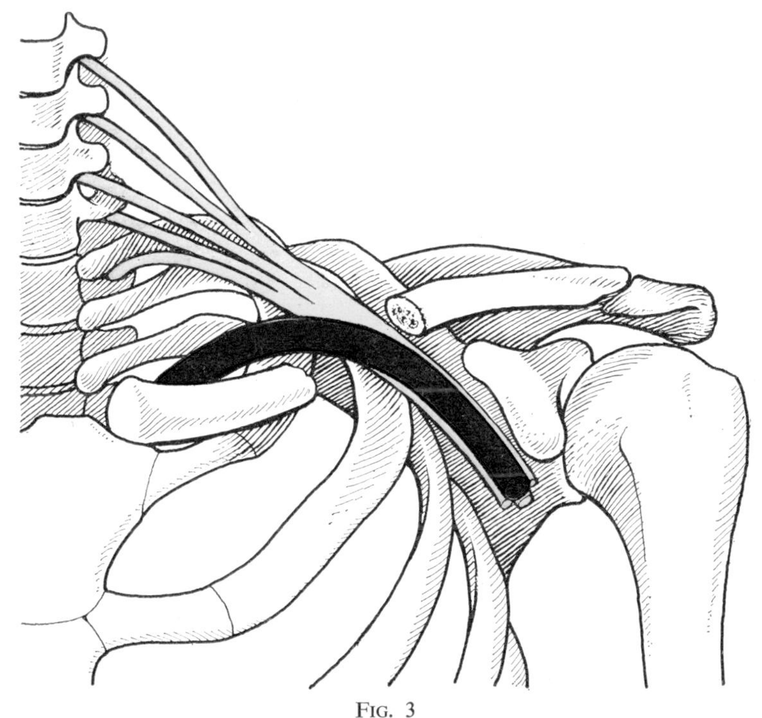

FIG. 3
The clavicle has been cut across to expose the nerves passing behind the artery as it enters the axilla.

namely the subscapularis, latissimus dorsi and teres major from above downwards. The apex of the axilla lies between the first rib medially, the clavicle in front and the upper border of the scapula behind. The floor is formed by fascia continuous with the deep fascia of the chest wall.

DISSECTION

In the upper part of the axilla, the axillary vessels and the infraclavicular part of the brachial plexus lie on the medial wall and lower down they cross to the lateral wall. The vessels and nerves should now be cleaned. Begin with the **axillary vein,** the most medial structure, whose many and variable branches should be removed. Follow the vein downwards to its formation by the junction of the **basilic vein** (the medial of the two main veins draining the superficial tissues of the upper limb) and the **brachial veins** (the veins accompanying the main artery of the upper arm). This junction usually takes place about the middle of the upper arm (Fig. 12). The axillary vein if followed upwards is seen to disappear deep to the clavicle and the cephalic vein joins it at about this level. It becomes the subclavian vein at the outer border of the first rib. Many of the tributaries of the axillary vein correspond to the larger branches of the artery. The vein lies medial to the artery.

The axillary vein should be divided about its middle and turned upwards and downwards. The **axillary artery** can now be seen, surrounded by the brachial plexus. The artery begins as a continuation of the subclavian artery at the outer border of the first rib and ends at the lower edge of the teres major where it continues as the brachial artery (Fig. 2). The branches which pass to the anterior axillary wall have been cut while reflecting the pectoral muscles. The **acromiothoracic artery** arises high up in the axilla and its branches pierce the clavipectoral fascia (page 12) and were found in the infraclavicular fossa. Passing to the posterior wall from the lower part of the axillary artery is the large **subscapular artery.** The **posterior** and **anterior humeral circumflex arteries** are found winding round the upper end of the humerus.

Now examine and clean the nerves. In this dissection only that part of the plexus below the clavicle can be easily exposed and the roots, trunks and divisions of the plexus will be dissected with the neck. The part of the brachial plexus above the clavicle (the trunks) lies above the subclavian artery. As the plexus and artery pass into the axilla, part of the plexus passes behind and then medial to the artery so that the lateral, posterior and medial cords come to have the relationship to the artery implied by their names. The terminal branches of the plexus are determined in the following way. The **median nerve** is formed on the lateral side of the artery by the union of two branches (the medial and lateral heads of the median nerve). The lateral head leads back to the **lateral cord** of the plexus from which the **musculocutaneous nerve** also arises. This nerve can be followed distally into the coracobrachialis muscle. The medial head of the median nerve can be followed anterior to the

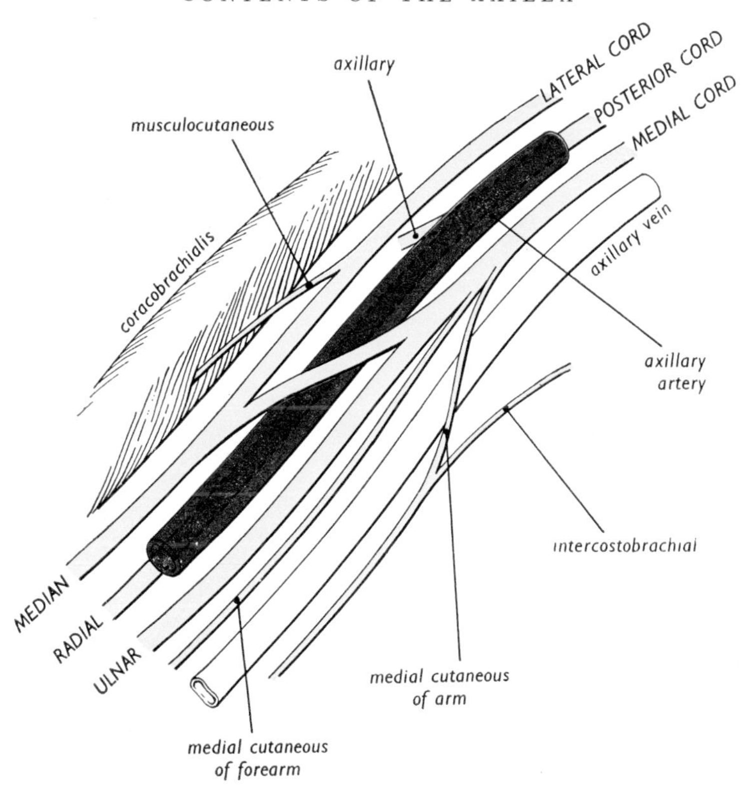

FIG. 4
Diagram of the cords and branches of the brachial plexus.

artery to the **medial cord.** The other large nerve arising from this cord is the **ulnar nerve.** Two sensory nerves also arise from the medial cord, the **medial cutaneous nerves of the arm** and **of the forearm.** These three nerves lie medial to the artery. The ulnar nerve can be followed along the medial side of the axillary artery and its continuation the brachial artery to about the middle of the upper arm, where the nerve passes backwards. The **posterior cord,** lying behind the artery, divides high up into the **radial nerve,** continuing downwards into the upper arm behind the artery, and the **axillary nerve** which winds backwards through an intermuscular space, reaches the deep surface of the deltoid and supplies it (Fig. 2). As the nerve passes backwards it lies close to the medial side of the surgical neck of the humerus and just below the capsule of the shoulder joint. Thus fracture of this part of the humerus or dislocation of the head downwards may injure the axillary nerve. Just before it divides, the posterior cord of the plexus gives off three branches to the three muscles forming the posterior wall of the axilla. The **lateral** and **medial pectoral nerves** will also be found arising from the lateral and medial cords respectively.

The posterior relations of the brachial plexus and artery should now be examined, although only the humeral attachments of the muscles forming this wall can be seen in the present dissection (Fig. 2). These are the subscapularis, latissimus dorsi and teres major from above downwards. Clean the fascia from the surface of the subscapularis and trace its tendon towards the humerus. The nerves to the subscapularis, latissimus dorsi and teres major lie in this fascia and should be traced to their destinations. Separate the latissimus dorsi from the teres major. Note that the latissimus dorsi winds round the inferior border of the teres major and comes to lie in front. Trace the tendon of latissimus dorsi to the humerus and then cut across it about 4 cm from its insertion. Reflect the cut ends and clean the teres major. In the gap between the teres major and subscapularis and deep to both, find the tendon of the **long head of the triceps brachii** as it passes upwards to the infraglenoid tubercle of the scapula. Follow the axillary nerve and the posterior circumflex humeral artery backwards between the subscapularis above and teres major below (Fig. 2). Note that these structures lie immediately below the capsule of the shoulder joint as they pass backwards. The **subscapularis** is attached laterally to the lesser tuberosity of the humerus and the capsule of the shoulder joint, the **latissimus dorsi** to the floor of the intertubercular groove and the **teres major** to the medial lip of this groove. All these muscles are supplied by branches of the posterior cord of the brachial plexus.

STRUCTURAL DETAILS

The axillary artery (Fig. 2)

There are many branches of the axillary artery supplying: (1) the thoracic wall (the medial wall of the axilla), (2) the pectoral muscles (the anterior wall of the axilla, (3) the shoulder and

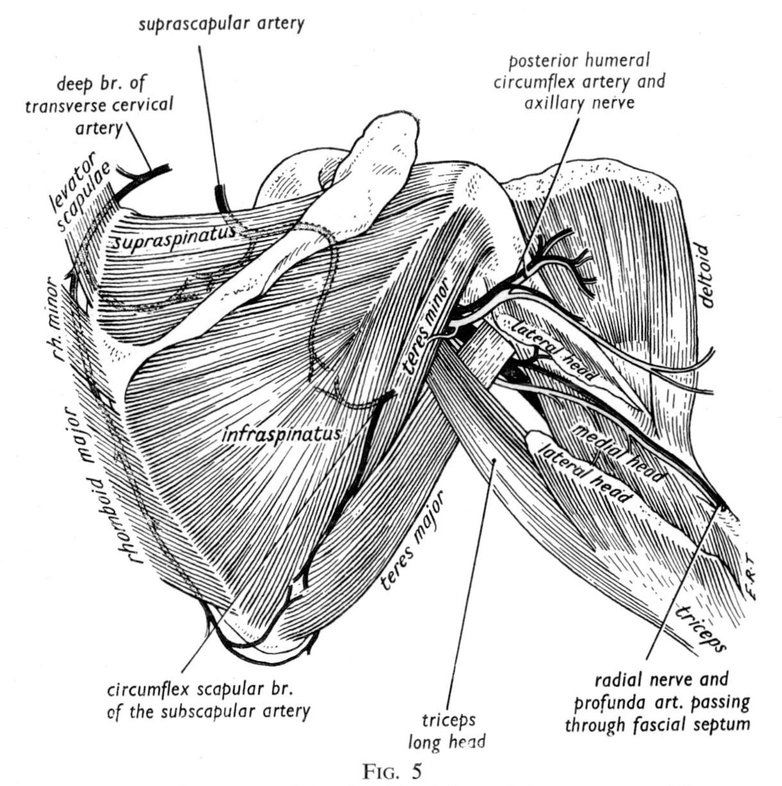

FIG. 5
The posterior aspect of the shoulder joint and the upper arm. The anastomosis round the scapula is indicated.

acromioclavicular joints, (4) the posterior wall of the axilla, (5) the shoulder region deep to the deltoid muscle. Of the branches, the largest are the **subscapular artery,** lying on the posterior axillary wall, and the **posterior circumflex humeral artery,** accompanying the axillary nerve through an intermuscular space to the deep aspect of the deltoid. The subscapular artery gives off a large branch, the **circumflex scapular artery,** passing backwards through another more medial intermuscular space and anastomosing with the branches of the subclavian artery in the region of the scapula (Fig. 5). Of the smaller axillary branches, the **acromiothoracic artery** is fairly constant and has a wide distribution to the anterior chest wall and the shoulder.

The brachial plexus (Figs. 2 and 6)

The brachial plexus provides a means by which fibres in various spinal nerves are re-arranged on their way to the periphery of the limb. The spinal nerves contributing to the plexus and the arrangement of the nerve roots may vary in different individuals and on the two sides of the same individual.

The ventral rami of the spinal nerves (usually C5 to T1) constitute the **roots** of the plexus and arise in the neck opposite the lower cervical vertebrae. The 5th and 6th cervical nerves form the **upper trunk,** the 7th the **middle trunk** and the 8th cervical and 1st thoracic nerves the **lower trunk.** These trunks are above the clavicle and each divides into an anterior and a posterior division behind the clavicle. Those nerve fibres destined to supply the dorsal (extensor) aspect of the limb now pass through the three **posterior divisions** into the **posterior cord.** Muscular and cutaneous branches of this cord pass in the subscapular, the thoracodorsal, the axillary and the radial nerves, and supply posterior structures from the shoulder to the fingers. Those nerve fibres destined to supply the ventral (flexor) aspect of the limb run in the three **anterior divisions** and form the **medial** and **lateral cords.**

The anterior divisions of the upper and middle trunks form the **lateral cord** and the anterior division of the lower trunk forms the **medial cord.** Branches of the lateral cord are distributed to many muscles of the front of the upper arm and to the skin of the lateral surface of the whole limb. The main branches of the lateral cord are the musculocutaneous and pectoral nerves and the lateral head of the median nerve. Branches of the medial cord are distributed to muscles of the front of the forearm and hand and to the skin of the medial surface of the whole limb. The main branches of the medial cord are the ulnar nerve, the medial head of the median nerve, pectoral nerve and the medial cutaneous nerves of the arm and forearm.

The **axillary nerve** supplies the deltoid and teres minor and has sensory branches to the back of the shoulder region and upper arm, and to the posterior aspect of the capsule of the shoulder joint. The **radial nerve** gives off cutaneous, muscular and articular branches in the arm, forearm and hand which supply posterior structures. The muscles producing extension at the elbow, wrist and fingers are all supplied by the radial nerve or its branches. The **musculocutaneous**

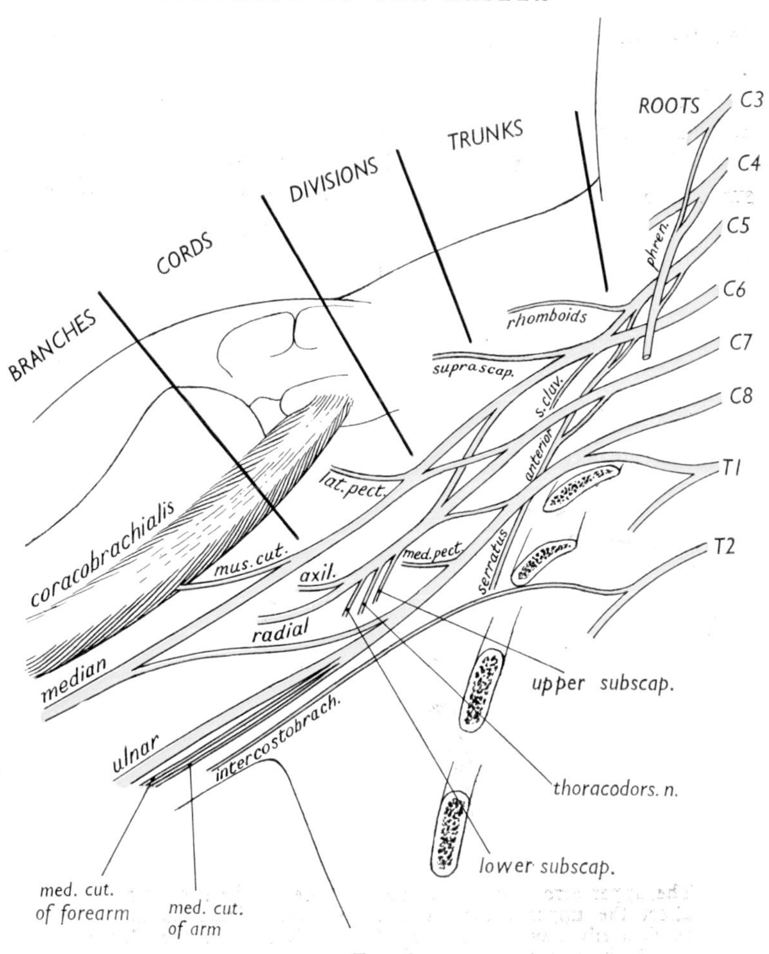

FIG. 6
The brachial plexus. For description, see text, page 22.

nerve pierces and supplies the coracobrachialis, supplies the biceps and brachialis and becomes cutaneous at the region of the elbow, supplying the skin on the lateral aspect of the forearm. The lateral head of the median nerve joins the medial head to form the **median nerve,** which lies at first lateral to the artery and then crosses in front to its medial side about the middle of the upper arm. The median nerve supplies most of the muscles on the front of the forearm and some of the muscles of the hand, and supplies the skin on the lateral side of the palm and the palmar surface of the lateral digits. Its terminal branches also supply the region of the nail and the finger tip.

The **ulnar nerve** has no branches in the upper arm but supplies some muscles of the forearm and most of the muscles of the palm of the hand. Its cutaneous distribution is to the palmar and dorsal surfaces of the medial side of the hand and the medial digits. The

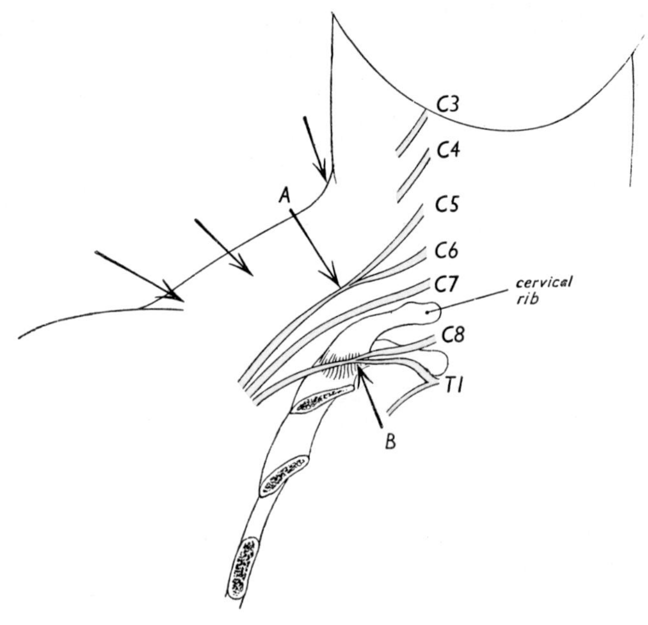

FIG. 7
The upper arrows indicate how pressure on the shoulder will affect the upper trunk. The lower arrow indicates how a cervical rib may stretch the 1st thoracic root of the plexus. Both these injuries give rise to characteristic paralyses.

medial cutaneous nerve of the arm with the intercostobrachial nerve supplies the medial aspect of both surfaces of the upper arm. The **medial cutaneous nerve of the forearm** supplies the medial aspect of both surfaces of the forearm. Anteriorly its distribution overlaps that of the lateral cutaneous nerve of the forearm (a branch of the musculocutaneous nerve) but posteriorly it is separated from the branches of this nerve by an area supplied by the posterior cutaneous nerve of the forearm from the radial nerve.

FUNCTIONAL ASPECTS

Though the blood supply to the upper limb is through the axillary artery, this may occasionally be blocked and alternative routes may be opened up. It is mainly through the scapular anastomosis that the circulation of the upper limb is maintained if the axillary artery is tied above the level of the subscapular artery (Fig. 5). If the axillary artery is tied near its origin, that is above the thoracic branches, the upper limb circulation may be maintained through anastomoses between the intercostal and internal thoracic arteries and the axillary thoracic branches, as well as through the scapular anastomosis.

Injuries to the brachial plexus are not uncommon and usually occur at two sites (Fig. 7). (1) The upper trunk [A] may be injured when the shoulder is depressed excessively. As a result, muscles and skin in the distribution of the 5th and 6th cervical roots are affected. The arm is held adducted and medially rotated at the shoulder, and extended and pronated at the elbow. (2) In the presence of a cervical rib, or if the 1st rib is pulled upwards, the 1st thoracic root may become stretched [B]. As a result, the small muscles of the hand are paralysed and the hand assumes a claw-like shape.

The lymph nodes in the axilla are very important because of their relation to malignant disease of the breast. They will be discussed in the section on lymph drainage (Chapter 12).

CHAPTER 5

THE POSTERIOR ASPECT OF THE SHOULDER GIRDLE

INTRODUCTION

IN this section, the main concern will be with the muscles which move the scapula on the chest wall. Many of the muscles to be dissected work against gravity and help raise the arm above the shoulder. The upper limb drops to the side of the body because of the action of gravity. This movement is controlled by the relaxation of the antagonist muscles.

If the back muscles have been dissected, turn to page 30.

Place the body on its face, with the shoulders raised on a block. Palpate the following bony landmarks on the cadaver and skeleton: the external occipital protuberance on the back of the base of the skull and in line with the spinous processes of the vertebrae, the superior nuchal line running laterally from the protuberance, and the spinous processes of the vertebrae down the midline, although those of the upper cervical vertebrae cannot be easily distinguished. Either the spinous process of the seventh cervical or that of the first thoracic vertebra is the most prominent. Palpate the spine and the inferior angle of the scapula (opposite the spinous process of the 3rd and 7th thoracic vertebrae respectively), the acromion which is the continuation laterally and forwards of the spine of the scapula, and the highest points of the iliac crests (on a line between the 3rd and 4th lumbar spinous processes).

DISSECTION

Incise the skin along the following lines: (1) from the external occipital protuberance laterally along the superior nuchal line to the ear; (2) from this point downwards along the lateral border of the neck to join the anterior incision along the clavicle at the tip of the shoulder; (3) down the midline of the back to the level of the highest point of the iliac crest; (4) from the lowest

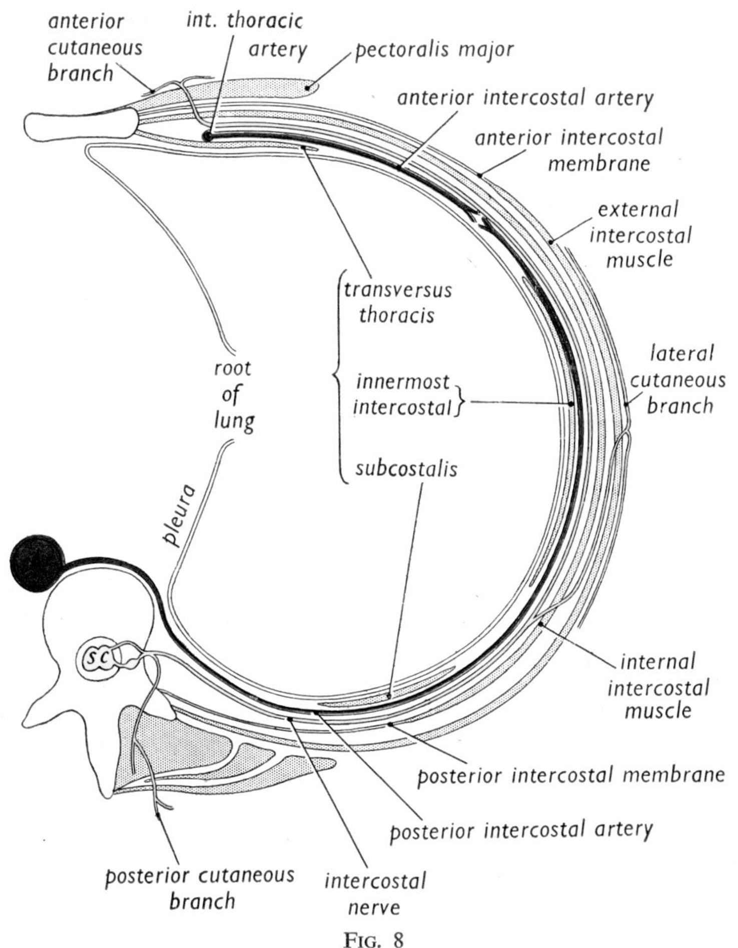

Fig. 8
Diagram of a section along an intercostal space showing the relations of the vessels and nerves to the muscular planes.

point of this line to the lowest point of the costal margin in the midaxillary line. Reflect this flap of skin downwards and laterally so that the whole of the thorax and upper half of the upper arm is denuded.

While reflecting the skin, some cutaneous nerves near the midline will be cut. These are branches of the dorsal rami of the spinal nerves (Fig. 8). The superficial fascia is often fairly dense with loculated fat, especially over the neck and lower part of the back. This should be removed, together with the deep fascia. The latter is continuous with the deep fascia over the front of the neck, the arm, the axilla, the chest and the abdomen, and is attached to those parts of the bones that are subcutaneous, namely the spinous processes of the vertebrae, the spine and acromion of the scapula and the crest of the iliac bone. In the midline of the neck, attached to the external occipital protuberance and the spinous processes of the cervical vertebrae, is a condensation of fibrous tissue called the ligamentum nuchae. Expose the **trapezius muscle,** passing from the midline (the occiput to the lowest thoracic spinous process) to the angle formed by the clavicle, acromion and spine of the scapula (Fig. 9). The lower part of the trapezius covers some of the **latissimus dorsi** as the latter passes from the vertebral spinous processes (6th thoracic to the lowest lumbar), sacrum and iliac crest to the upper end of the humerus. As far as possible define the attachments of these two muscles.

Cut the trapezius vertically 3 cm from the midline and turn it laterally, taking care not to damage the deeper muscles. Entering its deep surface from the neck is the spinal accessory nerve together with the transverse cervical artery, which is a branch of the subclavian artery and takes part in the scapular anastomosis. By pulling the shoulder forwards, the scapula is pulled away from the midline and a layer of muscles attached to its medial border is more easily seen.

Attached to the superior angle of the scapula is the **levator scapulae,** passing upwards into the neck where it is attached to the transverse processes of the upper cervical vertebrae (Fig. 9). As its name suggests this muscle raises the scapula upwards on the chest wall. If the scapula is fixed one levator scapulae pulls the neck laterally, and both pull the neck backwards.

The **rhomboid minor** and **major** are attached to the lower part of the ligamentum nuchae and to the spinous processes of the upper thoracic vertebrae and pass downwards and laterally to the medial border of the scapula (Fig. 9). The minor lies above the major. When they contract they pull the scapula upwards and medially and also rotate the bone about an anteroposterior axis so that the

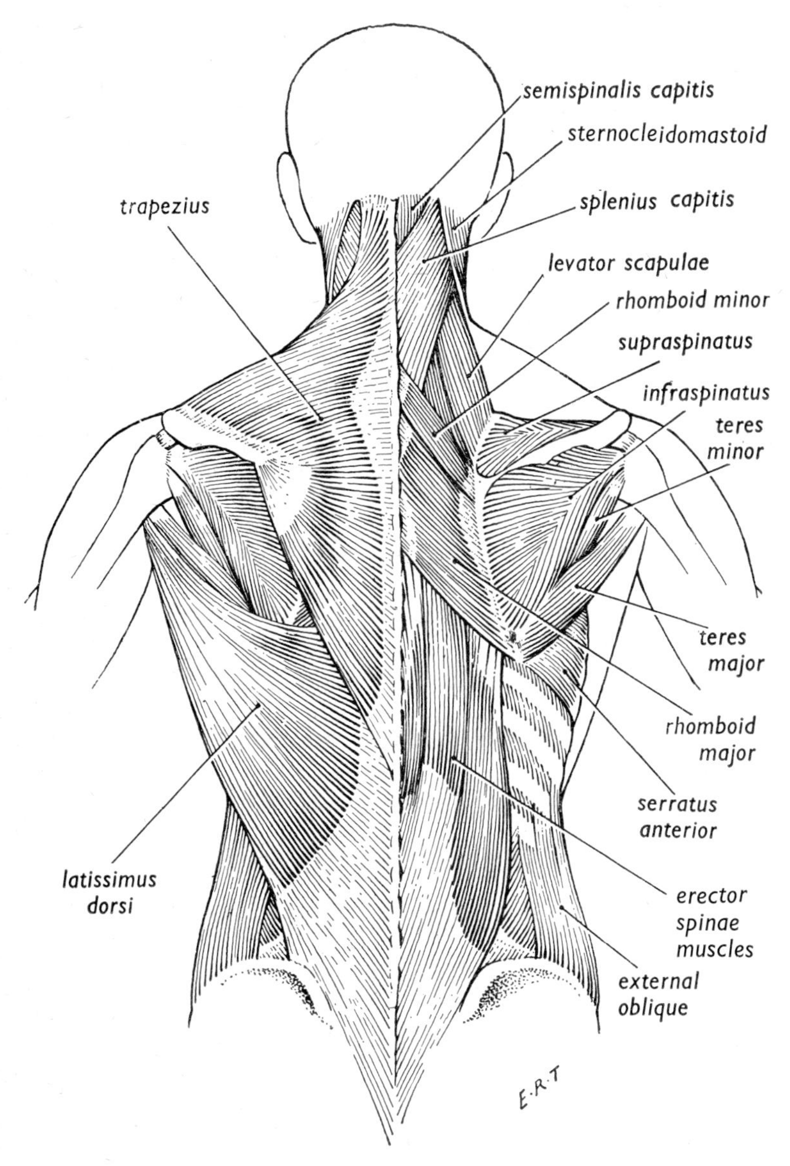

FIG. 9

The muscles of the back. On the right side, the trapezius and latissimus dorsi have been removed.

glenoid cavity looks downwards. Detach the rhomboid muscles from the spinous processes, turn them laterally and so expose the nerves to the muscles.

STRUCTURAL DETAILS

The scapula

The main part of this bone is a triangular plate lying on the back of the chest wall over an area extending from the 2nd to the 7th ribs. The anterior (costal) surface, next to the ribs, is flat, but the posterior surface has a horizontal shelf, the **spine,** projecting backwards and dividing this surface into an upper quarter, the **supraspinous fossa** and a lower three-quarters, the **infraspinous fossa.** The spine when followed medially gradually disappears and when followed laterally turns forward to form a flattened process of bone, the **acromion** which has an upper and lower surface and a medial and lateral edge. The medial edge of the acromion articulates with the lateral end of the clavicle.

The scapula has superior, inferior and lateral angles, the last carrying the **glenoid cavity** for articulation with the head of the humerus. The glenoid cavity in the anatomical position faces forwards and laterally. The upper border of the scapula has a notch **(suprascapular notch)** and between the notch and glenoid cavity is the **coracoid process,** a piece of bone projecting forwards and laterally for about 3 cm. The lateral border of the scapula extends downwards and medially from the glenoid cavity to the inferior angle and the medial border vertically from the superior to the inferior angle.

The scapula **ossifies** in cartilage from a considerable number of centres. The primary centre appears in the body at about the end of the 2nd month of fetal life (Fig. 10). At birth the coracoid process, the acromion, the glenoid cavity, the medial border and inferior angle are still cartilaginous. Secondary centres appear in most of these areas about puberty and fuse with the body by the 25th year. The centre for the coracoid process appears during the 1st year and fuses at puberty.

The scapula has attached to it many of the muscles by which the limb is moved and is itself attached to the trunk by further

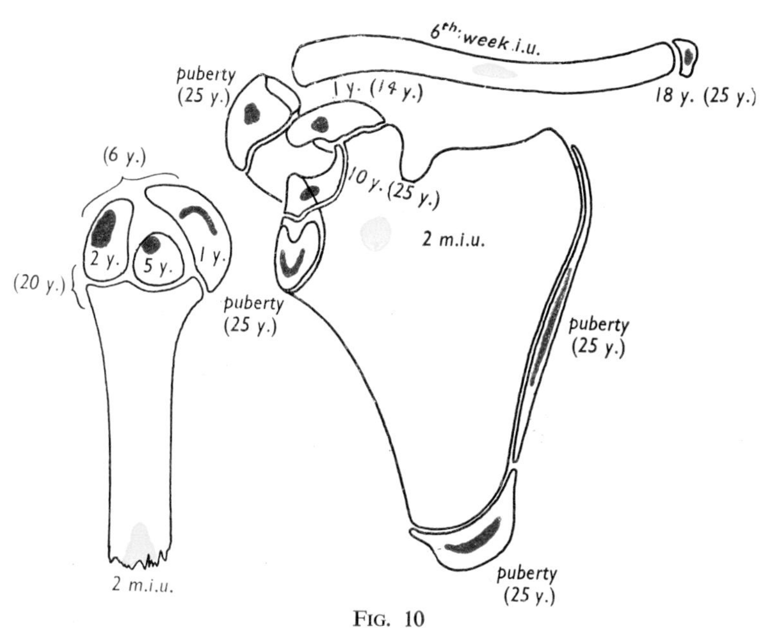

FIG. 10

The primary centres of ossification, shown in yellow, appear in the 2nd month of intra-uterine life. The secondary centres, shown in red, appear in a long bone (the humerus) between the 1st and 5th years and in a flat bone (the scapula) mainly about puberty. The times of fusion, shown in brackets, vary from 18 years to 25 years (m.i.u, months in utero; y, years).

muscles. The weight of the limb and of objects held by it is transferred to the vertebral column through the scapula and clavicle. The scapula is attached to the lateral end of the clavicle by the strong coracoclavicular ligaments. When the arm is raised laterally, the scapula rotates about an anteroposterior axis at this point of suspension from the clavicle. This rotation turns the glenoid cavity upwards and thus adds to the large angle of movement of the upper limb on the trunk and increases the usefulness of the limb. The scapula can also be moved forwards round the chest wall as in movements involving pushing with the limb, and upwards as in shrugging the shoulders.

The humerus

This is a typical long bone, the upper end of which has a rounded **head,** facing medially and slightly backwards, for articulation with the glenoid cavity. The articular area of the head is much larger than that of the glenoid cavity. The head is joined to the **body** by the **anatomical neck.** The upper end of the **body** anteriorly has a vertical depression called the **intertubercular groove** (with a floor and lateral and medial lips). By determining the position of the head and intertubercular groove, the upper end and medial and anterior aspects of the bone can be decided, and this serves to define the side of the body to which a humerus belongs. Medial to the upper end of the intertubercular groove is the **lesser tuberosity,** and lateral to the upper end of the groove and projecting laterally is the **greater tuberosity.** The part of the body immediately below the tuberosities is the **surgical neck;** it is a common site of fracture. Laterally, about the middle of the body is a triangular roughened area, the **deltoid tuberosity.** Posteriorly, running downwards from the medial to the lateral side of the body along its middle third is the **radial groove.**

Ossification of the body of the humerus begins in a primary centre at about the end of the 2nd month of intra-uterine life (Fig. 10). Secondary centres for the upper end appear for the head, for the greater tuberosity and for the lesser tuberosity between one and five years. These centres fuse together at six years and join the body at twenty years. The humerus increases in length more at the upper epiphyseal plate than at the lower.

The trapezius

The trapezius is attached to the structures of the midline—the external occipital protuberance, the ligamentum nuchae and the spinous processes of the thoracic vertebrae. The upper fibres pass downwards, the middle horizontally and the lower upwards, all converging towards their lateral attachment to the arch of bone formed by the upper edge of the spine of the scapula, the medial edge of the acromion and the posterior surface of the lateral third of the clavicle. The trapezius is supplied by the spinal accessory nerve. The nerve enters the deep surface of the muscle, passing under its upper edge from the side of the neck.

The trapezius of each side, acting as a whole, pulls the scapula towards the midline of the back, as in bracing the shoulders. The upper fibres acting alone elevate the scapula, as in shrugging the shoulders. The lower fibres, which are attached to the spine of the scapula, help the serratus anterior to rotate the scapula so that the glenoid cavity looks upwards. (This will be considered more fully with the movements at the shoulder joint.) Both muscles acting together pull the head backwards, *i.e.* extension of the head on the neck. One trapezius alone pulls the head backwards and turns the face to the opposite side.

The latissimus dorsi

The latissimus dorsi is attached to the spinous processes of the lower thoracic vertebrae and indirectly to the lumbar and sacral vertebrae by the **thoracolumbar fascia.** It is also attached to the posterior part of the crest of the iliac bone and to the lower ribs. Its fibres pass upwards and laterally, converging towards their tendinous attachment on the floor of the intertubercular groove. Some of the fibres arise from the inferior angle of the scapula where it lies under the upper edge of the muscle. The tendon winds round the lower border of the teres major and twists on itself. At its humeral attachment, the muscle forms part of the posterior wall of the axilla. Its nerve (thoracodorsal) comes from the posterior cord of the brachial plexus. The latissimus dorsi extends and medially rotates the humerus. If a hammer is held above the head with the elbow straight and is wielded from that position downwards, the force of the stroke is increased by the action of the latissimus dorsi. In climbing with the arms this muscle draws the body upwards.

CHAPTER 6

THE SHOULDER JOINT AND ITS STABILISING MUSCLES

INTRODUCTION

THE shoulder joint is surrounded by a number of muscles that stabilise the joint because of their close proximity and attachment to the capsule, and also move the humerus in many directions.

On the skeleton and the cadaver, identify the lateral part of the clavicle, the acromion, the spine of the scapula, the greater and lesser tuberosities of the humerus and the deltoid tuberosity.

DISSECTION

Examine the **deltoid,** which lies over the shoulder joint and extends down the outer aspect of the upper arm. It is attached above to the arch of bone formed by the lower border of the spine of the scapula, the outer edge of the acromion and the anterior surface of the lateral third of the clavicle. Its fibres converge downwards and the muscle ends in a tendon attached to the deltoid tuberosity of the humerus. Tendinous intersections can be seen in the vertical part of this muscle and the fibres in this part are short, passing obliquely from one intersection to another, so that a multipennate muscle is formed. The muscle is supplied by the axillary nerve and though its main action is abduction, the anterior fibres flex and medially rotate the upper arm and the posterior fibres extend and laterally rotate it at the shoulder joint.

Detach the deltoid from the clavicle and scapula and turn it downwards. Immediately deep to its upper attachment is the **subacromial bursa,** a synovial sac, separating the deltoid from the superior aspect of the shoulder joint capsule and the tendons attached to it. This bursa extends medially beneath the acromion, the coracoid process and the ligament joining these two processes.

While turning the deltoid downwards find the axillary nerve entering its deep surface posteriorly. This nerve passes backwards from the axilla, between the transversely running teres minor above and teres major below. Immediately lateral to the nerve is the body of the humerus and immediately medial is the long head of the triceps. The space through which the axillary nerve passes is known as the quadrilateral space (Figs. 2 and 5). From in front, the upper boundary of this space is the subscapularis muscle, and between the subscapularis and the teres minor behind is the capsule of the shoulder joint. Downward dislocation of the humerus (the commonest type of dislocation) frequently injures the axillary nerve. The axillary nerve also supplies teres minor and gives off the **upper lateral cutaneous nerve of the arm,** which supplies the skin over the lower two-thirds of the deltoid.

Examine the supraspinatus in the supraspinous fossa, the infraspinatus in the infraspinous fossa, and the teres minor and teres major attached to the lateral border of the scapula (Fig. 5). Clear definition of the supraspinatus and infraspinatus muscle fibres is difficult because of the thick fascia covering them and providing attachment for a large number of their fibres. The **supraspinatus** is attached to the supraspinous fossa and passes laterally under the acromion, from which it is separated by the subacromial bursa. The **infraspinatus** is attached to the infraspinous fossa and the **teres minor** is attached to the middle of the lateral border of the scapula. All three muscles pass laterally and end in tendons attached to the upper, middle and lower facets on the greater tuberosity of the humerus. All three tendons are adherent to the capsule of the shoulder joint. The lateral attachments of the infraspinatus and teres minor can only be separated from each other with difficulty. They are lateral rotators at the shoulder joint. The supraspinatus is an abductor. In order to see teres major clearly it may be necessary to remove the upper fibres of the latissimus dorsi from the region of the inferior angle of the scapula. The **teres major** is attached to the inferior angle of the scapula and passes laterally to end in a tendon attached on the front of the humerus to the medial lip of the intertubercular groove. It forms part of the posterior wall of the axilla. Its nerve supply from the posterior cord of the brachial plexus has already been seen. Its actions are similar to those of latissimus dorsi; it extends and medially rotates the humerus.

Divide the supraspinatus vertically across its middle and look for the suprascapular vessels and nerve entering its deep surface from in front and above. Follow them round the lateral edge of the spine of the scapula into the infraspinous fossa, where they supply the infraspinatus (Fig. 5). Starting at the medial border of the scapula, detach the fibres of the infraspinatus muscle from the bone as far as the lateral border. Identify the circumflex scapular branch of the subscapular artery entering the infraspinous fossa from in front at the level of the middle of the lateral border and anastomosing with the suprascapular artery (a branch of the subclavian artery) in the infraspinous fossa. The suprascapular nerve comes from the upper trunk of the brachial plexus (Fig. 6) and its origin can be seen in the dissection of the neck. It crosses the upper border of the scapula in the suprascapular notch and like the artery, passes from the supraspinous fossa into the infraspinous fossa. Thus the axillary and suprascapular nerves are concerned with abduction and lateral rotation of the humerus. Both nerves receive their fibres from the 5th and 6th cervical nerves.

Divide the levator scapulae close to its scapular attachment. Attached to the upper border of the scapula medial to the suprascapular notch is the **inferior belly of the omohyoid,** which should be divided at its attachment to the scapula. Divide the suprascapular nerve and vessels just before they enter the supraspinous fossa and the attachments of the trapezius to the scapula and the clavicle.

Turn the body on to its back and elevate the medial part of the clavicle. Lateral to the sternoclavicular joint is the **costoclavicular ligament,** attached to the clavicle above and to the first costal cartilage below.

Examine the **acromioclavicular joint** by cutting through its capsule all round the joint, and observe the direction of the joint surfaces. The plane of this joint is anteroposterior and oblique from above downwards in a medial direction with the result that the clavicle dislocates upwards and outwards at this joint. This is prevented by a very strong ligament passing from the inferior surface of the lateral end of the clavicle to the coracoid process lying just below it, the **coracoclavicular ligament.**

Tie two ligatures round the axillary artery and the brachial plexus as near to the clavicle as possible. Cut between the ligatures. Divide the serratus anterior behind its nerve and divide any remaining vessels or nerves passing from the trunk to the limb. Remove the upper limb from the trunk. Although most of the clavicle has been left attached to the body it must be considered as part of the shoulder girdle, and functionally it plays an important part in the movements of the upper limb on the trunk.

It is now possible to study more closely the attachment of the serratus anterior to the inferior angle and medial border of the scapula. The attachment of the **subscapularis** to a series of ridges on the costal surface of the scapula can be demonstrated by cutting across the muscle along the line of the lateral border of the scapula and turning the muscle medially. The lateral, more tendinous part of the muscle is firmly attached to the capsule of the shoulder joint but an attempt should be made to trace the muscle to the lesser tuberosity of the humerus. The subscapularis is a medial rotator of the humerus.

In order to see the **capsular ligament** it is necessary to remove the muscles that are firmly attached to it, the supraspinatus above, the infraspinatus and teres minor behind and the subscapularis in front (Figs. 2 and 5). The capsule is attached firmly to the bony circumference of the glenoid cavity medial to the **glenoidal labrum** which is a fibrocartilaginous ring attached to the edge of the articular surface. Laterally the capsule is attached to the anatomical neck of the humerus, except inferiorly, where it is attached to the body of the bone about 2 cm below the neck. The capsule is loose, especially inferiorly. It is thickened where it bridges the intertubercular groove to form the **transverse humeral ligament,** and in front to form the **glenohumeral** and **coracohumeral ligaments.**

The tendon of the **long head of the biceps** is exposed by opening the capsule superiorly by an incision from the edge of the glenoid cavity to the intertubercular groove. This tendon is attached to the upper edge of the glenoid cavity and glenoidal labrum and passes over the head of the humerus inside the capsule. It emerges and lies in the intertubercular groove between the pectoralis major laterally and the latissimus dorsi medially. Observe the arch above the joint, formed by the acromion, the coracoid process and the coraco-acromial ligament.

Open the joint completely by a circular incision and examine the **synovial membrane** lining the capsule as far as the articular cartilage. The tendon of biceps is surrounded by a sheath of synovial membrane as it passes laterally inside the capsule. Beneath the tendon of subscapularis there is a bursa communicating with the joint. Examine the glenoidal labrum which is covered by synovial membrane on its inner and outer aspects.

STRUCTURAL DETAILS

The shoulder joint

The shoulder joint is the main joint whereby the upper limb moves on the trunk but the range of movement is considerably increased by the movements of the scapula and clavicle. The head of the humerus and the glenoid cavity form a ball-and-socket joint. The articular area on the head of the humerus is much greater than that of the glenoid cavity. In any position of the joint only a small part of the head is in contact with the glenoid cavity. This also means that the bony surfaces make little contribution to the stability of the joint, although the glenoid cavity is somewhat deepened by the glenoidal labrum. The blood supply to the joint comes from the various vessels in the region, namely the suprascapular, subscapular and circumflex humeral arteries. Its nerve supply is derived from the suprascapular, axillary and subscapular branches of the brachial plexus. This follows the general principle that the nerves supplying the muscles moving a joint also supply the joint capsule and the skin over the joint.

FUNCTIONAL ASPECTS

The movements of the shoulder

When the upper limb moves on the trunk, in addition to the humerus moving on the scapula, the clavicle and scapula move on the chest wall and on each other. At the shoulder joint itself abduction, adduction, flexion, extension, circumduction and medial and lateral rotation occur. Abduction, adduction, flexion and extension of the upper limb are increased by the movements of the pectoral girdle on the chest wall. The scapula can move upwards, downwards, forwards and backwards. It can rotate on the chest wall so that the glenoid cavity looks upwards as the inferior angle of the scapula moves laterally. This movement is often referred to as **lateral rotation** of the scapula. The opposite movement is called **medial rotation.** When the scapula rotates laterally it usually moves forwards on the chest wall and when it rotates medially it usually moves backwards.

Movements of the scapula are accompanied by movements of the clavicle. The fulcrum for the movements of the clavicle is the costoclavicular ligament which is very near the sternoclavicular joint. When the scapula moves upwards the lateral end of the clavicle moves upwards with it, and the medial end moves downwards, but to a much lesser extent. Similarly, when the scapula moves forwards, the lateral end of the clavicle moves with it and the medial end moves slightly backwards. The forward movement is produced by contraction of serratus anterior and results in a considerable increase in the reach of the limb (Fig. 11).

When the scapula rotates, the fulcrum is the coracoclavicular ligament, so that there is considerable movement at the acromioclavicular joint. The lateral end of the clavicle moves upwards to some extent with lateral rotation of the scapula and downwards with medial rotation and the clavicle also rotates about its long axis.

The main muscles moving the scapula from the anatomical position: (1) upwards are the trapezius and levator scapulae, (2) downwards are the inferior fibres of trapezius and the latissimus dorsi, (3) forwards are the serratus anterior and pectoralis minor, (4) backwards are the rhomboids and trapezius, (5) so that the glenoid cavity looks upwards are serratus anterior and the lower part of trapezius, (6) so that the glenoid cavity looks downwards are the pectoralis minor and the rhomboids.

When testing the movements of the upper limb on the trunk the terms abduction, adduction, flexion and extension refer to movements of the limb in the coronal and sagittal planes of the body. It should be understood, however, that the plane of the shoulder joint is not sagittal, but lies obliquely so that the head of the humerus in the anatomical position faces backwards as well as medially and the plane of the joint passes backwards and laterally. Most anatomical movements in the coronal and sagittal planes produce some torsion of the capsule that would not occur if the

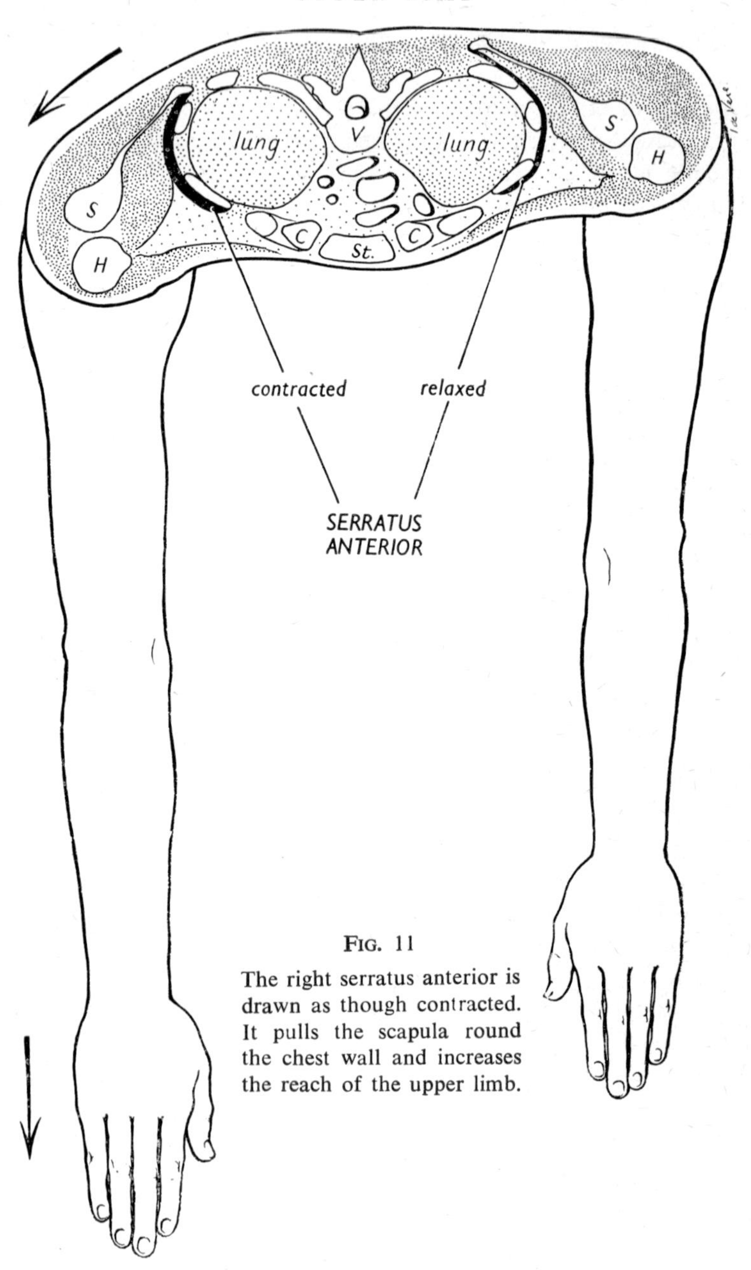

FIG. 11

The right serratus anterior is drawn as though contracted. It pulls the scapula round the chest wall and increases the reach of the upper limb.

movements were related to the plane of the glenoid cavity. Thus in anatomical flexion there is also a certain amount of abduction from the scapular plane.

Abduction of the upper limb from the trunk can be carried out until the limb is raised above the head, that is, the limb can move outwards in the coronal plane through 180°. In the initiation of abduction, the supraspinatus is used. Its main action is to hold the head of the humerus close to the glenoid cavity, so that the deltoid instead of pulling the humerus straight up, causes the humerus to move round an anteroposterior axis through the head. After about 20° of abduction the deltoid can continue to act without the help of the supraspinatus. The tendon of the supraspinatus is liable to rupture. If this occurs, abduction of the upper limb from the position of hanging alongside the body is difficult, but if the limb is passively abducted to about 20° the deltoid can then take over the rest of the movement. Full abduction of the upper limb is produced by abduction of the humerus at the shoulder joint and lateral rotation of the scapula. These movements occur simultaneously after about 50° of abduction. Glenohumeral movement accounts for about 120° of the total abduction and scapular rotation for about 60°. The main muscles producing lateral rotation of the scapula are the serratus anterior and the lower fibres of the trapezius. Some lateral rotation of the humerus is an essential part of abduction beyond 90°.

Adduction of the upper limb from the anatomical position is prevented by the trunk, but the limb can be brought inwards towards the midline, close to and across the chest wall for about 30°. The main adductor is pectoralis major. The chief factors controlling the lowering of the upper limb from the position of 90° of abduction are gravity and the controlled relaxation of the deltoid.

The upper limb can be **flexed** through 180°. Beyond 90°, flexion involves lateral rotation of the scapula and medial rotation of the humerus. The chief flexors are the anterior fibres of the deltoid, the clavicular fibres of the pectoralis major, and the coracobrachialis.

Extension is limited to about 30°. The main extensors are the latissimus dorsi and teres major.

The upper limb can be **rotated** laterally at the glenohumeral joint through about 60°. Medial rotation is limited by the trunk to about 30° when the elbow is flexed. When the elbow is extended medial rotation may be confused with pronation of the forearm. If the middle of the back can be touched from above and from below, full lateral and medial rotation are present. The main lateral rotators of the humerus are the infraspinatus, teres minor and the posterior fibres of the deltoid. The chief medial rotators are the subscapularis, teres major and latissimus dorsi.

The shoulder joint is more frequently dislocated than any other joint in the body. The bony surfaces make only a small contribution towards its stability, and its capsule is loose and has no strong ligaments associated with it. The tendons attached to the capsule and to the tuberosities help to stabilise the joint, except inferiorly, where dislocation most frequently takes place. In addition, the long head of the biceps gives some support superiorly, and the acromion, coracoid process and coraco-acromial ligament act as a secondary socket for the head of the humerus. The deltoid also protects the joint above, in front and behind. Dislocation most frequently takes place when the humerus is abducted, in which position the head of the humerus lies on the inferior part of the capsule.

CHAPTER 7

THE UPPER ARM AND ELBOW REGION

INTRODUCTION

THE articulating surface of the lower end of the humerus has two areas of very different shapes, the pulley-like **trochlea** for the ulna medially and the rounded **capitulum** for the radius laterally. These allow two types of characteristic movement from the anatomical position, flexion and pronation of the forearm.

It is clear that very different mechanisms must come into play in each of them. In flexion besides movement of the ulna on the trochlea, the radius must also move on the capitulum. But in pronation the radius moves on the capitulum, rotating round the more or less fixed axis of the ulna. For the latter movement there are joints between the radius and ulna, the superior and inferior radio-ulnar joints. When the upper limb is in the anatomical position note that the long axis of the forearm is not in line with the long axis of the upper arm. The lateral deviation of the forearm is called the **carrying angle,** which is measured on the lateral side of the elbow.

On the skeleton and cadaver identify the **epicondyles** of the humerus projecting medially and laterally above the elbow. At the back of the elbow is the **olecranon** of the ulna. The **coronoid process** of the ulna can be seen on the skeleton in front but cannot be identified on the cadaver.

DISSECTION

Carry the end of the anterior incision made in the dissection of the axilla straight down the arm to a point 8 cm distal to the humeral epicondyles and from here transversely round the forearm. Remove the whole cuff of skin marked out, taking care not to damage vessels and nerves in the subcutaneous tissues. A subcutaneous bursa may be found over the back of the olecranon.

The cephalic vein, already identified in the dissection of the axilla, should be traced distally as it lies in the superficial fascia.

On the medial side of the upper arm, in its lower half, find the **basilic vein** lying in the superficial fascia. Clean the vein which, when followed proximally, passes through the deep fascia at about the level of the middle of the upper arm (Fig. 12). Dissect out the cephalic and basilic veins in front of the elbow joint and note any cutaneous nerves which are present. At this level an obliquely placed **median cubital vein** generally runs from one to the other. This short, wide vein lies close to the surface and is frequently used for intravenous injections. There is much variation in the pattern of the veins. Examine the arrangement of the veins in the cubital fossa of your own limbs and note that they may be different on the two sides.

Remove the superficial fascia and cut through the deep fascia by longitudinal midline incisions in front and behind. Turn the flaps of fascia medially and laterally, preserving any cutaneous nerve trunks. It will be noticed that on the lateral and medial sides of the arm the fascia is continuous with fascial sheets passing into the interior of the limb. These are the **lateral** and **medial intermuscular septa,** which, below, are attached to the **lateral** and **medial supracondylar ridges** of the humerus and divide the limb into anterior (flexor) and posterior (extensor) compartments.

The flexor compartment

The deeper structures of the upper arm are now partially exposed. On the front of the upper arm the **biceps brachii** is seen (Fig. 1). Its long and short heads have been already identified; trace them downwards and clean the belly of the muscle as far as the cubital fossa, where it forms a tendon. A strong band of fascia (the **bicipital aponeurosis**) leaves the tendon and, passing superficially and medially, ends in the deep fascia of the forearm (Fig. 13). Identify this tendon and aponeurosis on your own arm.

Behind the medial side of the short head of the biceps identify the **coracobrachialis muscle** and clean it down to its attachment to the middle of the medial side of the humerus. Identify and trace the musculocutaneous nerve through the substance of the coracobrachialis on to the deep surface of the biceps.

Cut through the biceps at the junction of its upper and middle thirds and turn the ends downwards and upwards. Clean the musculocutaneous nerve lying deep to the muscle and trace it to

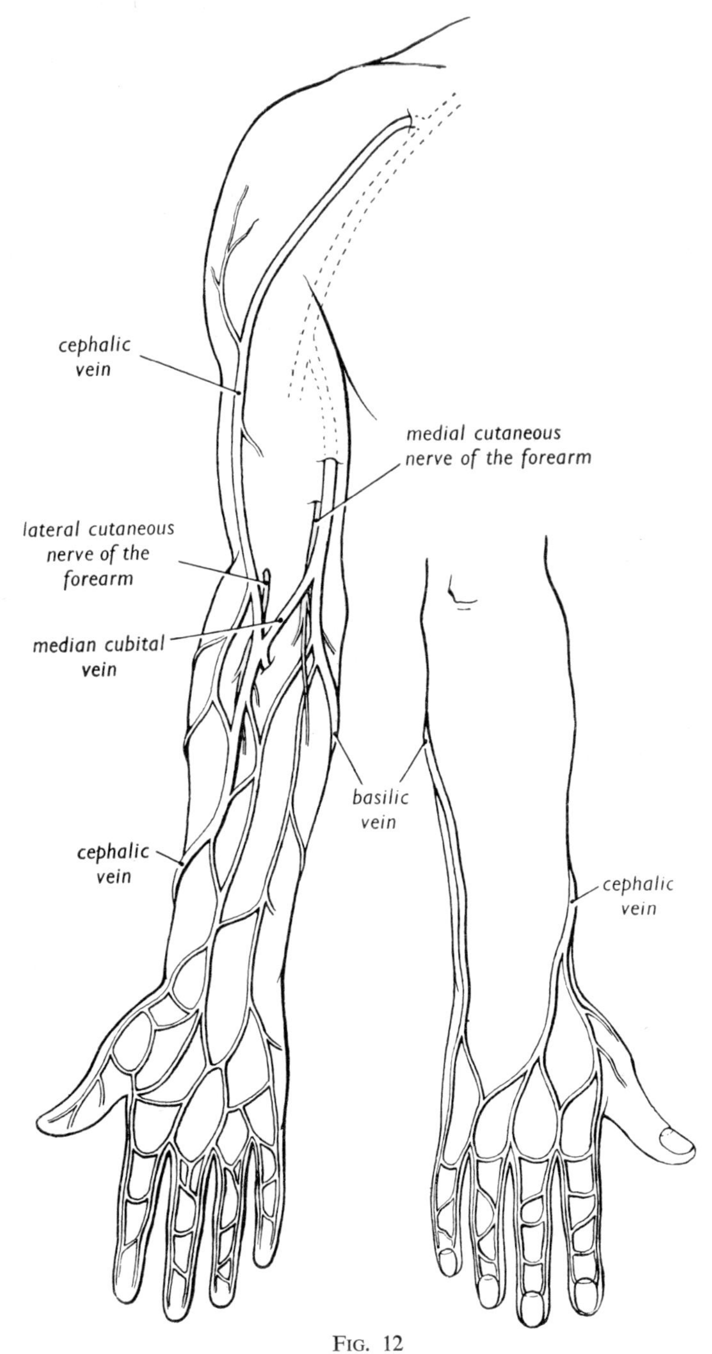

FIG. 12

The superficial veins of the upper limb. Note the close proximity of the cutaneous nerves to the veins in the cubital fossa.

the lateral side of the limb. It gives branches to the muscles of the front of the upper arm. Trace the nerve to the point just above the lateral side of the elbow where it pierces the deep fascia to become the **lateral cutaneous nerve of the forearm,** which is distributed to the skin of the forearm above the wrist (Fig. 24).

Behind the biceps muscle lies the **brachialis,** arising from the front of the lower half of the humerus. Clean its surface down to the level of the cubital fossa. Its tendon is attached to the front of the coronoid process of the ulna.

Medial to the biceps find the **brachial artery** and **veins.** Lower down they enter the cubital fossa approximately halfway between the epicondyles, in front of the brachialis muscle and medial to the tendon of the biceps. The brachial artery is a continuation of the axillary artery. It may be covered only by skin and fascia along its whole length but it is usually overlapped by the medial edge of the biceps. In the cubital fossa it is covered anteriorly by the bicipital aponeurosis. Above, the artery lies on the insertion of the coracobrachialis and below on the brachialis. The biceps is lateral throughout the whole length of the artery. The median nerve crosses it anteriorly from the lateral to the medial side. The ulnar nerve is medial and the radial nerve posterior to the artery in the first part of its course in the upper arm. Identify the largest branch of the brachial artery, the **profunda brachii** (Figs. 2 and 5). This comes off the main artery in the proximal part of the upper arm and passes backwards and laterally into the extensor compartment with the radial nerve, where it lies in the radial groove. The terminal branches of the profunda artery pass distally and take part in the anastomosis round the elbow joint, supplying branches to its lateral side. The brachial veins, if followed up, are joined by the basilic vein and form the axillary vein (Fig. 12).

Find the **median nerve,** which in the upper part of the upper arm lies lateral to the brachial artery but about halfway down crosses in front of the artery and then lies medial to it. The **ulnar nerve** lies medial to the artery in the upper part of its course but halfway down the arm it passes backwards through the medial intermuscular septum. Trace it down behind this septum deep to the medial head of the triceps to the level of the elbow joint and verify that at this level it lies subcutaneously, behind the medial epicondyle. Medial to the ulnar nerve in the upper part of the arm are the **medial cutaneous nerve of the arm** and **the medial cutaneous nerve of the forearm.** Trace the latter down to the level of the cubital fossa, where it pierces the deep fascia to become subcutaneous.

The extensor compartment

On the back of the upper arm remove the deep fascia completely. Clean the posterior surface of the **triceps muscle** and examine its proximal attachments (Fig. 5). Above and medially, find its **long head,** arising from the scapula below the glenoid cavity. Clean this head and trace it to its union with the **lateral head,** attached to the upper part of the back of the humerus above the radial groove, and the **medial head,** attached to the posterior surface below the radial groove. The muscle is attached by a tendon to the upper surface of the olecranon. Many of its fibres are attached to the deep fascia on the back of the elbow. The muscle is an extensor of the forearm at the elbow joint. The triceps is supplied by the radial nerve.

Find the **radial nerve,** lying between the long head of the triceps and the body of the humerus. Cut downwards and laterally, through the lateral head of triceps just above the middle of the upper arm, taking care not to injure the radial nerve (Fig. 5). In this part of its course the radial nerve, as it lies against the bone, may become involved in fractures of the humerus. Trace the nerve upwards and downwards and find its cutaneous branches and the branches for the three heads of the triceps. When traced upwards the radial nerve is seen to arise from the posterior cord of the brachial plexus and to lie posterior to the axillary and brachial arteries. It enters the radial groove and lies between the lateral and medial heads of the triceps. Distally, beyond the radial groove, the nerve pierces the lateral intermuscular septum and enters the flexor compartment. Find the nerve as it lies in front of the septum on the brachialis, and is overlapped laterally by the brachioradialis (Fig. 13).

The cubital fossa

This fossa lies in front of the elbow and is triangular in shape with the apex distally. The base is formed by a line joining the two epicondyles. Identify the muscles forming its sides—the

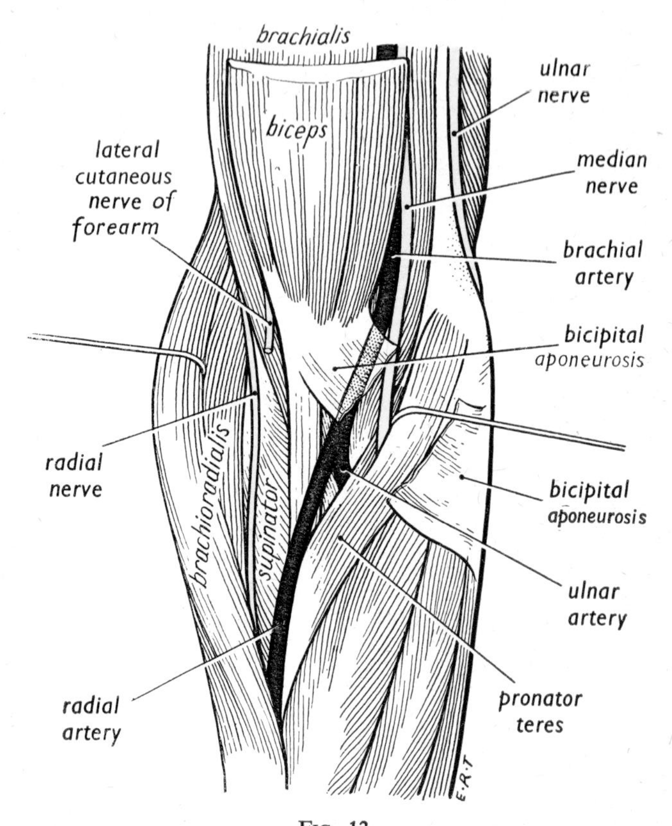

FIG. 13
The cubital fossa dissected to show the important structures.

brachioradialis laterally and above, and the **pronator teres** medially (Fig. 13). Clear away the fascia forming its roof and follow the brachial artery and the median nerve as far as the apex of the fossa. Remove all the small arteries and veins and find the branch of the nerve to the pronator teres. At the apex of the fossa, or just proximal to it, the brachial artery divides into the **radial artery** laterally, and the **ulnar artery** medially. The radial artery should be followed to the apex of the fossa under cover of the edge of the brachioradialis muscle. At this point it meets the radial nerve which lies on its lateral side. The ulnar artery passes more deeply, leaving the fossa deep to the pronator teres muscle. The **common interosseous artery** arises just below the beginning of the ulnar artery and is easily found.

The median nerve leaves the fossa by passing through the pronator teres muscle. It gives a branch to the pronator teres in the fossa and usually at the same level gives off the **anterior interosseous nerve,** which leaves the fossa with the median nerve and passes to the deep muscles of the forearm.

Pull aside the brachioradialis and find the radial nerve running along its deep surface. It supplies the brachioradialis and extensor carpi radialis longus. Follow the nerve until it gives off a large branch, the **posterior interosseous nerve,** which passes into the substance of the supinator on the lateral aspect of the upper end of the radius in the floor of the fossa. Medial to the supinator examine the brachialis and its tendon, which form the floor of the fossa above and medially. Trace the tendon of the biceps to its attachment on the tuberosity of the radius. After pronating the forearm pull on the biceps tendon and note its supinating action.

STRUCTURAL DETAILS

The lower end of the humerus

The body of the humerus in transverse section is circular above and roughly triangular below with anteromedial, anterolateral and posterior surfaces. The lateral and medial borders, when traced downwards become sharper and are known in this region as the **medial** and **lateral supracondylar ridges.** They end below as small elevations, the **medial** and **lateral epicondyles.** Above the articular edge of the humerus anteriorly are two hollows, the **coronoid fossa** medially and the **radial fossa** laterally, into which fit the prominences of the coronoid process of the ulna and the head of the radius respectively when the elbow is flexed.

The posterior surface of the humerus is smooth. It ends below at the upper margin of the **olecranon fossa,** lying immediately above the articular surface and receiving the olecranon of the ulna when the elbow is extended.

The articular surface of the lower end of the humerus has two regions, a lateral rounded **capitulum** with which the head of the radius articulates, and a medial, the **trochlea,** which articulates with the ulna. The ulnar and humeral surfaces fit very closely and their shape allows only movements of flexion and extension, as in a hinge. The capitulum and the head of the radius do not fit

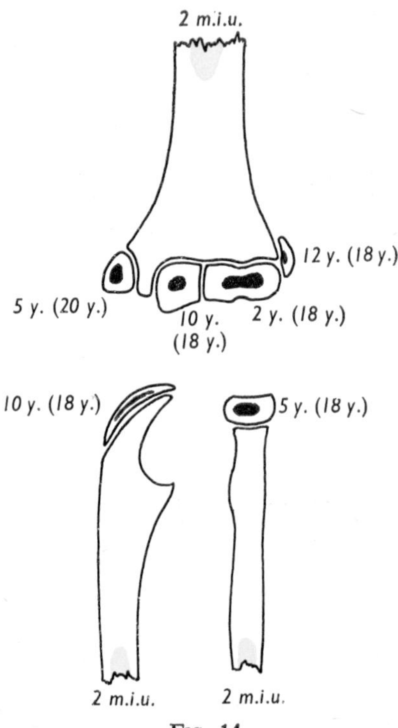

FIG. 14

The primary centres of ossification are shown in yellow and the secondary centres in red. The figures in brackets indicate the times of fusion of primary and secondary centres (m.i.u, months in utero; y., years).

together and this allows the rotational movements of **pronation** and **supination** as well as the hinge movements. The medial epicondyle of the humerus is larger and more prominent than the lateral epicondyle. Both can be felt beneath the skin and posteriorly the medial ridge of the trochlea limits a groove lateral to the medial epicondyle. In this groove lies the ulnar nerve.

Secondary centres of **ossification** for the lower end of the humerus appear in the capitulum, the medial epicondyle, the trochlea and the lateral epicondyle between two and twelve years. The capitulum, trochlea and lateral epicondyle fuse together by 14 years and join the body at 18 years. The medial epicondyle remains detached from the body in the male until twenty years. A narrow wedge of the body, on which lies the ulnar nerve, separates the medial epicondyle from the trochlea until fusion is complete (Fig. 14).

The upper end of the radius

The upper end of the radius has a **head** joined to the **body** by a **neck.** The head is cylindrical and entirely covered by articular cartilage; its upper surface is slightly concave. The neck is constricted and at its junction with the body on the medial side lies the **radial tuberosity.** Passing downwards and laterally from the tuberosity is the oblique line that becomes the anterior border of the radius. Passing downwards on the medial side of the radius is the interosseous border.

The upper end of the ulna

On the upper end of the body of the ulna are two forward projections, an upper the **olecranon,** and a lower the **coronoid process.** Between the two is a smooth articular surface, the **trochlear notch,** for articulation with the trochlea of the humerus. On the lateral side of the coronoid process is the concave **radial notch** for articulation with the head of the radius. From the posterior margin of this facet a well-marked ridge, the **supinator crest,** runs down for 3 - 4 cm on the lateral side of the shaft of the bone to become continuous with the interosseous border.

Both the radius and ulna have a primary centre of **ossification** in the body, appearing at about the end of the second month of intra-uterine life (Fig. 14). Secondary centres appear for the head of the radius and the top of the olecranon at the fifth and tenth year respectively and join the bodies at the eighteenth year.

The flexor muscles

The **biceps** is attached by a short head to the tip of the coracoid process and by a long head to the scapula just above the glenoid cavity and to the glenoidal labrum. The long head at its origin and for the first 5 cm of its course lies within the capsule of the shoulder joint and is surrounded by a thin sleeve of synovial membrane. It lies directly above the head of the humerus and leaves the joint by running in the intertubercular groove beyond which it becomes muscular and joins the short head. The muscle is attached distally by a strong tendon to the posterior part of the tuberosity of the radius. In addition, a strong band of fibrous tissue, the **bicipital aponeurosis,** leaves the medial side of the tendon and becomes continuous with the deep fascia over the flexor muscles of the forearm medially. The main actions of the biceps are flexion at the elbow joint and supination of the fore-arm. This latter action is due to its attachment to the back of the radial tuberosity. When the forearm is pronated (palm facing back-wards), the tendon passes posteriorly between the radius and ulna. Contraction of the biceps pulls on the radial tuberosity so that the radius rotates on its long axis and the forearm is supinated (palm facing forwards). The long head, by virtue of its relation to the upper end of the humerus, may exert a stabilising influence on the joint by steadying the head of the bone and prevent its displace-ment upwards.

The **brachialis** acts only on the humero-ulnar joint and the **coracobrachialis** only on the shoulder joint. All these flexor muscles are supplied by branches of the musculocutaneous nerve. Some of the forearm muscles assist flexion at the elbow joint and will be examined later.

The extensor muscles

The origin of the **triceps** has been examined and the action of the long head in supporting the inferior aspect of the shoulder joint noted. The lateral and medial heads are attached to the posterior surface of the humerus, above and below the radial groove respec-tively (Fig. 5). The three heads join and form a common attach-ment to the posterior part of the superior surface of the olecranon. The short **anconeus,** passing from the back of the lateral epicondyle to the upper end of the ulna behind the supinator crest, also extends the elbow. The extensor muscles are supplied by the radial nerve.

The veins (Fig. 12)

Usually there are two main superficial veins of the arm, the **cephalic** laterally and the **basilic** medially. They arise from the sides of a dorsal venous arch on the back of the hand, which receives branches from the fingers and hand. From the lateral side of the venous arch the cephalic vein runs up the lateral border of the forearm, then in the furrow lateral to the biceps and finally in the cleft between the deltoid and the pectoralis major muscles. It ends by piercing the clavipectoral fascia and joining the axillary vein. The basilic vein arises from the medial side of the dorsal arch, runs up the medial border of the forearm, and then in the furrow medial to the biceps. It pierces the deep fascia in the middle of the upper arm and joins the brachial veins. The valves in these superficial veins can be demonstrated in the living subject. Lying along the side of the cephalic vein is the lateral cutaneous nerve of the forearm and along the basilic vein is the medial cutaneous nerve of the forearm. These nerves are usually deep to the veins in the cubital fossa.

FUNCTIONAL ASPECTS

Note that the biceps tendon, the brachial artery and the median nerve lie in that order from the lateral to the medial side of the cubital fossa (Fig. 13). The cubital fossa is an important region of the upper limb since its superficial veins are used for intravenous injections and for obtaining specimens of blood for laboratory investigation, and the brachial artery in this part of its course is used for determining the blood pressure.

Although it would be possible to describe some of the structures of and movements at the elbow joint at this stage, consideration will be given to these after the upper attachments of the forearm muscles have been described (Chapter 9).

CHAPTER 8

THE FOREARM AND THE BACK OF THE HAND

INTRODUCTION

THE radius, ulna and carpus of man show in general a primitive pattern. The special features found in man all result in obtaining increased freedom of movement. The most striking specialisations are (1) only the radius articulates with the proximal row of carpal bones, (2) the movements of pronation and supination. In the supine position the two bones are parallel and in pronation the lower end of the radius, carrying the hand, moves over the lower end of the ulna, which remains relatively fixed. The upper end of the radius rotates within an annular ligament and the radius and ulna are held together by an interosseous membrane.

The main muscles producing supination are the biceps and the supinator. Pronation is performed mainly by proximal and distal muscles specialised for this purpose, pronator teres and pronator quadratus. The movements of pronation and supination are useful not only in actions such as turning a door handle or screwdriver but also for bringing the hand into the position required for holding a pen or a cup. The semiprone position places least strain on the muscles and the capsules of the joints.

On the cadaver and skeleton identify the lower ends of the radius and ulna. Each has a distal projection, the **styloid process.** The subcutaneous border of the ulna can be felt posteriorly from the olecranon to the styloid process. Distal to the radius and ulna are the carpal bones of the wrist, the metacarpal bones of the palm and the phalanges of the digits.

DISSECTION

Make incisions down the medial and lateral borders of the forearm and continue them into the hand as far as the thumb and little finger. Take care not to cut the structures lying subcutaneously as the skin is very thin in some regions, especially near the wrist. Remove the skin from the back of the limb.

In the superficial fascia over the back of the hand find the plexus of veins. Arising from this plexus and passing up the forearm are the cephalic and basilic veins, running respectively on the lateral and medial sides. Near the cephalic vein at the wrist, find the **radial nerve,** which supplies the lateral side of the back of the hand. Through its dorsal digital branches it supplies the skin of the back of the lateral three or three and a half fingers as far as, but not including, the terminal phalanges, which are supplied by the digital branches of the median nerve. The terminal part of the **dorsal branch of the ulnar nerve** is on the medial side of the back of the hand. The deep fascia is closely attached to the underlying muscles and over the back of the wrist it is thickened to form the **extensor retinaculum,** which is attached to the ridges on the posterior surface of the radius, to the styloid process of the ulna and to the medial side of the carpus. It holds the tendons in place. Carefully remove the deep fascia from the back of the forearm, but leave the retinaculum intact. Notice that the tendons in the region of the retinaculum are surrounded by a double layer of smooth tissue, the **synovial sheaths.**

The muscles of the back of the forearm are arranged in two groups, superficial and deep. Clean the superficial muscles and identify their attachments. Examine the tendons on the back of the hand, noting the interconnecting bands of fibrous tissue, and demonstrate them on your own hand. Carefully remove the skin from the back of the index and little fingers and note that each has two tendons. Clean the **dorsal interosseus muscles** lying between the metacarpal bones and note that each tendon is attached to the base of a proximal phalanx (see below) and to an extensor tendon (Figs. 15 and 19).

The superficial muscles of the back of the forearm (Fig. 15)

The **extensor carpi ulnaris, extensor digiti minimi, extensor digitorum** and **extensor carpi radialis brevis** all arise from the common extensor origin on the front of the lateral epicondyle of the humerus. The extensor carpi ulnaris has an additional origin from the posterior border of the ulna (middle two quarters). The **extensor carpi radialis longus** and the **brachioradialis** arise from the lateral supracondylar ridge, the latter muscle coming from

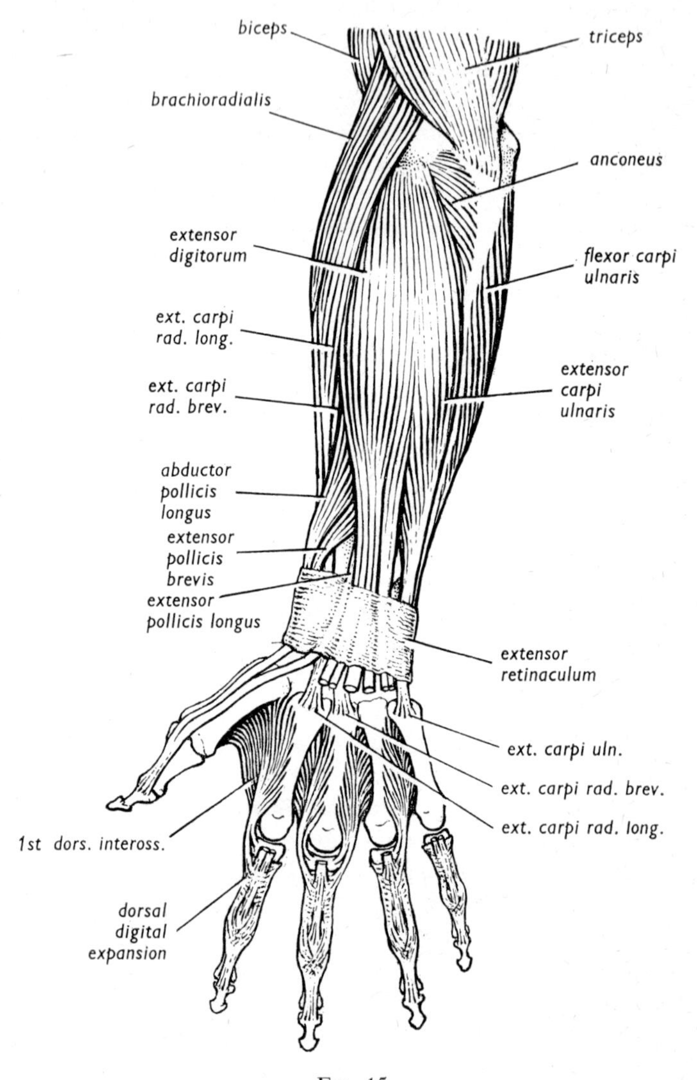

FIG. 15
The superficial muscles of the back of the forearm and hand.

the upper part of the ridge. The brachioradialis is attached to the lateral aspect of the lower end of the radius. The extensors of the wrist are attached to the bases of metacarpal bones, radialis longus to the 2nd, radialis brevis to the 3rd and ulnaris to the 5th. The long extensors of the fingers have a complicated insertion on to the backs of the phalanges. The tendons spread out over the metacarpo-phalangeal joints and the central portion runs distally to be attached to the base of the middle phalanx. The outer parts of the expansion have attached to them one or two interosseus tendons and a lumbrical tendon. These outer parts fuse together over the body of the middle phalanx and are attached to the base of the distal phalanx. The extensor digitorum is inserted on to the index, middle, ring and little fingers, the tendon of the extensor digiti minimi fusing with the digitorum tendon.

The deep muscles of the back of the forearm (Fig. 16)

Cut across the superficial muscles at their musculotendinous junctions and turn back the upper and lower parts to expose the deep group of muscles. Identify the transversely running fibres of the supinator passing from the ulna to the radius in the upper part of the forearm. Find the **posterior interosseous nerve** entering the region near the lower border of the supinator muscle which will be examined in detail later. This nerve supplies most of the muscles of the back of the forearm as it runs down between the superficial and the deep groups of muscles. The blood supply of these muscles comes from the anastomosis round the elbow joint and from the **posterior interosseous artery** (a branch of the common interosseous artery) which passes backwards between the two bones above the interosseous membrane and runs with the posterior interosseous nerve. They both end in the region of the back of the wrist joint.

Three of the deep group of muscles arise from the posterior surface of the radius or ulna and the interosseous membrane, and the most lateral and proximal, the **abductor pollicis longus,** comes from both bones. Identify the tendon of the abductor pollicis longus and medial to it the tendon of the extensor pollicis brevis. They lie in a groove on the lateral side of the radius, cross the wrist and are inserted into the base of the 1st metacarpal (abductor pollicis longus), and the base of the proximal phalanx (extensor pollicis

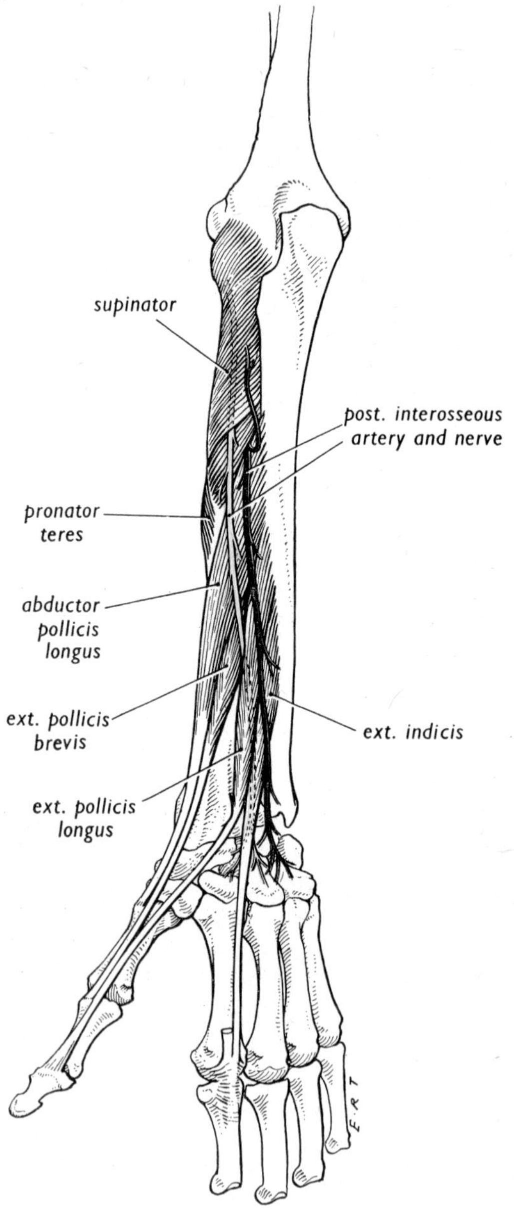

FIG. 16
The deep muscles of the back of the forearm.

brevis). Identify the extensor pollicis longus and medial to it the extensor indicis as they lie on the radius and cross the back of the wrist joint. The extensor pollicis longus runs in the deep groove on the medial side of the dorsal tubercle on the back of the lower end of the radius and is attached to the base of the distal phalanx of the thumb. The extensor indicis tendon fuses with the digitorum tendon on the back of the proximal phalanx of the index finger.

The "**anatomical snuff box**" lies on the lateral side of the wrist and is bounded laterally by the tendons of the abductor pollicis longus and extensor pollicis brevis and medially by the tendon of the extensor pollicis longus. In the floor of the hollow is the scaphoid bone on which lies the radial artery. It also contains branches of the radial nerve. Outline the hollow on your own hand and try to feel the pulsations of the artery.

The **dorsal interosseus muscles** are seen lying between the metacarpal bones (Fig. 19). The first is the largest, and is proximal to the web of the thumb. Identify this muscle on yourself. Each dorsal interosseus arises from the adjacent sides of two metacarpal bones. They act as abductors of the index, middle and ring fingers from an axis line running through the middle finger. Each tendon is inserted on to the side of a proximal phalanx and into the extensor expansion. The first tendon goes to the index finger, the second and third to the middle finger and the fourth to the ring finger.

The superficial muscles of the front of the forearm (Fig. 17)

Remove the skin from the front of the forearm and trace the cephalic and basilic veins to the region of the elbow. Certain cutaneous nerves will be found in the forearm. They are branches of the musculocutaneous nerve laterally and the medial cutaneous nerve of the forearm medially. The latter nerve becomes superficial above the elbow where the basilic vein pierces the deep fascia. Its anterior branch lies next to the vein in the forearm and its posterior branch winds backwards at the level of the elbow to be distributed to the medial side of the back of the forearm. The dorsal branch of the ulnar nerve comes off about 5 cm above the wrist and supplies the medial part of the back of the hand. It emerges from under the lateral border of the flexor carpi ulnaris.

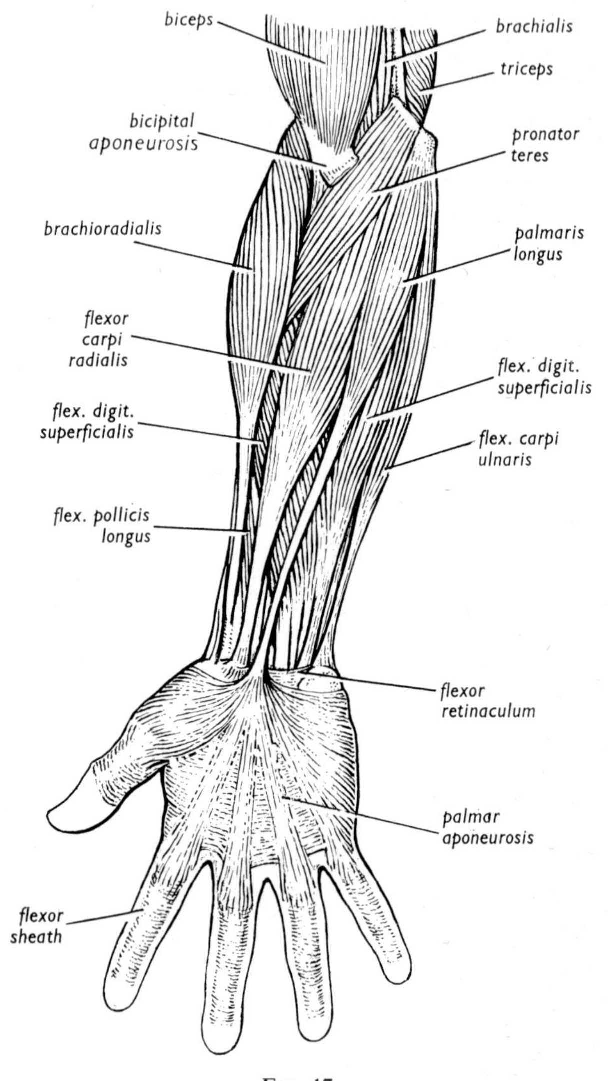

FIG. 17
The superficial muscles of the front of the elbow and forearm.

Remove the deep fascia on the muscles of the forearm and define the superficial group of four muscles, the **pronator teres, flexor carpi radialis, palmaris longus** and **flexor carpi ulnaris,** from lateral to medial. The palmaris longus may be absent. They are all attached to the front of the medial epicondyle, the **common flexor origin.** Pronator teres is also attached to the coronoid process of the ulna and flexor carpi ulnaris to the posterior border of the ulna. Follow pronator teres to the rough area on the middle of the lateral surface of the radius where it is attached. Follow the tendons of the other muscles to the wrist. Clean the brachioradialis muscle and, on its deep surface, find the **radial nerve.** It passes downwards on the muscles attached to the radius under cover of the brachioradialis, then backwards deep to its tendon, and becomes subcutaneous in the back of the forearm above the wrist. Its **posterior interosseous** branch arises in front of the lateral epicondyle, winds backwards through the substance of supinator muscle and was dissected on the back of the forearm. The radial artery lies on the medial side of the radial nerve in the lower two-thirds of the forearm.

The **median nerve** supplies muscular branches to all the muscles on the front of the forearm except the flexor carpi ulnaris and the medial half of the flexor digitorum profundus. The palmar cutaneous branch arises above the wrist and runs in the superficial fascia to supply the skin over the thenar eminence and the lateral side of the palm.

The **ulnar nerve** lies between the flexor carpi ulnaris (superficial) and flexor digitorum profundus (deep) and gives branches to both muscles. In its distal course in the forearm it lies subcutaneously on the lateral side of the flexor carpi ulnaris. The dorsal cutaneous branch has been seen passing to the back of the medial side of the hand and the medial fingers. A palmar cutaneous branch arises above the wrist, runs in the superficial fascia and supplies the medial side of the palm and the skin over the hypothenar muscles.

Clean the radial and ulnar arteries and preserve their main branches. The **radial artery** passes distally and laterally, and about one-third of the distance down the forearm meets the radial nerve under cover of the brachioradialis muscle. It lies on the

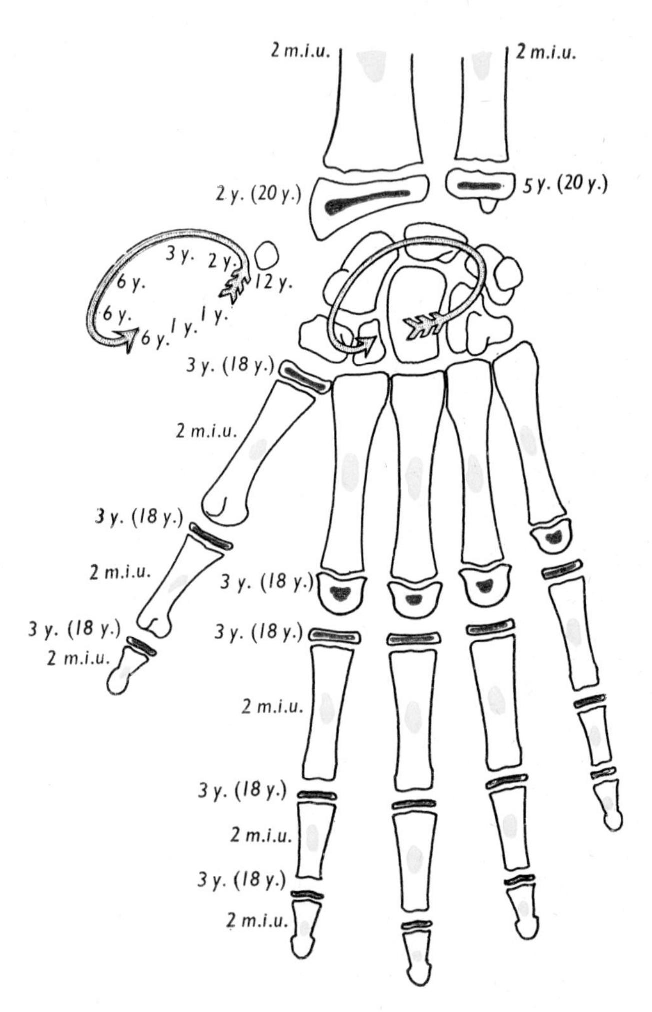

FIG. 18

The primary centres of ossification are shown in yellow and the secondary centres in red. The figures in brackets indicate the times of fusion of primary and secondary centres (m.i.u., months in utero ; y., years).

muscles attached to the front of the radius and then on the distal part of the anterior surface of the bone itself where it is subcutaneous and is easily palpated in the living subject. It gives off a recurrent branch to the elbow joint near its origin, and muscular, carpal and palmar branches. The **ulnar artery** passes medially and distally deep to the two heads of the pronator teres and then accompanies the ulnar nerve. It lies on the flexor digitorum profundus and in its upper course is covered by the flexor carpi ulnaris but above the wrist it is subcutaneous. Near its origin, it gives off recurrent branches to the elbow joint and the common interosseous artery which divides into anterior and posterior branches as well as numerous muscular branches.

Remove the skin from the front of the wrist and the palm of the hand thus exposing the **flexor retinaculum** and the **palmar aponeurosis.** Notice that the tendon of the palmaris longus (if present) is inserted into the palmar aponeurosis and is superficial to the retinaculum which is attached to the tuberosity of the scaphoid, the pisiform, the hook of the hamate and the tubercle of the trapezium (page 68). Distally the retinaculum is continuous with the palmar aponeurosis, which is especially thick over the middle of the palm. It also covers the thenar eminence (thumb muscles) laterally and the hypothenar eminence (little finger muscles) medially. Make a longitudinal incision through the middle of the flexor retinaculum and follow the tendon of the flexor carpi radialis to the bases of the 2nd and 3rd metacarpal bones. The flexor carpi ulnaris is attached to the pisiform.

Cut through the tendons of the flexor carpi ulnaris, flexor carpi radialis and palmaris longus 5 cm above the wrist and expose the **flexor digitorum superficialis,** forming the broad sheet of muscle over the deepest layer of muscles. It is attached to the common flexor origin on the medial epicondyle of the humerus, to the coronoid process of the ulna and to the oblique line on the front of the radius. Do not injure the radial and ulnar vessels lying near the muscle. The four tendons of flexor digitorum superficialis can be followed to the wrist where the two outer tendons lie deep to the two inner. Taking care not to injure the median nerve as it lies on the deep aspect of the muscle, cut through these tendons about 8 cm above the wrist and expose the deepest muscles. The nerve

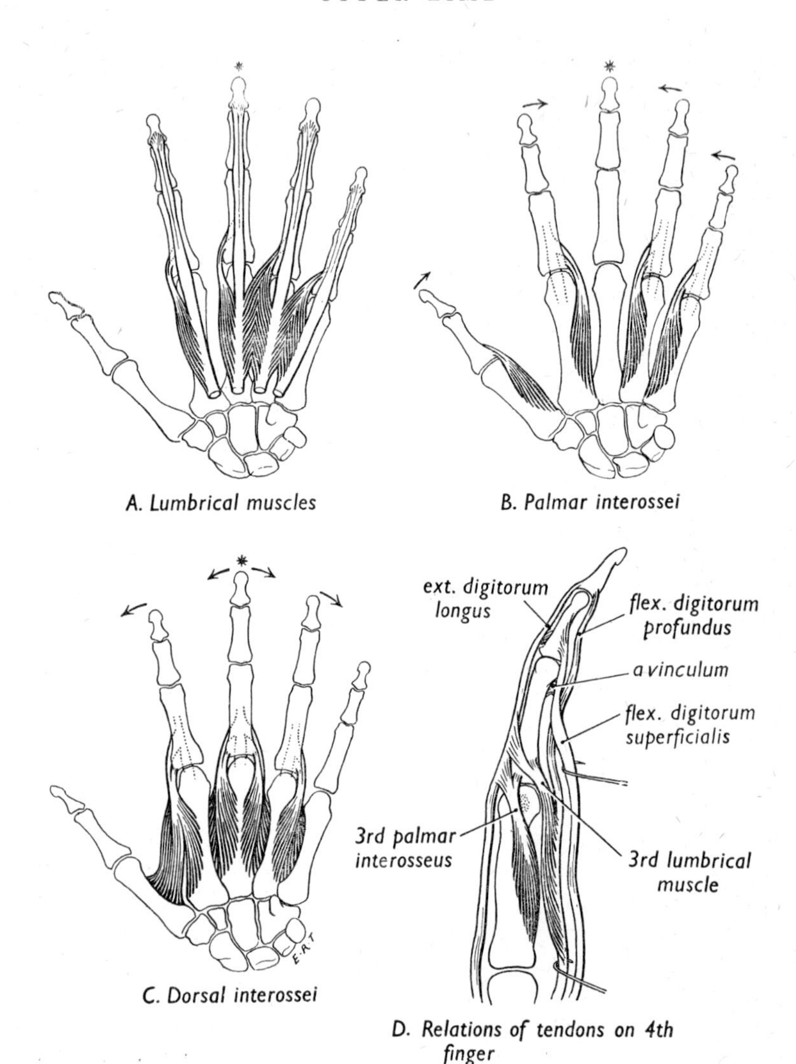

FIG. 19

The attachments of the lumbrical muscles (A), the palmar interossei (B), and the dorsal interossei (C). The * indicates the axis digit and the arrows indicate the actions of the muscles. In (D) an attempt has been made to show the attachments of the tendons of the palmar muscles to the radial side of the proximal phalanx and also to the extensor expansion of the ring finger.

passes deep to the flexor retinaculum and lies on the medial side of the tendon of the flexor carpi radialis.

The deep muscles of the front of the forearm (Fig. 20)

The deepest layer of muscles consists of the flexor pollicis longus, the flexor digitorum profundus, the pronator quadratus and the supinator.

So that the deepest structures can be more readily seen the pronator teres should be cut across just above its radial attachment. Clean the median and ulnar nerves and note their main branches. The **anterior interosseous** branch of the median nerve accompanies the anterior interosseous artery and lies in front of the interosseous membrane, between the flexor pollicis longus laterally and the flexor digitorum profundus medially. Make a longitudinal incision through the **pronator quadratus** as it lies across the front of the distal quarters of the radius and ulna and follow the nerve to the front of the wrist joint. It supplies the flexor pollicis longus, the lateral half of the flexor digitorum profundus and the pronator quadratus.

The deepest layer of muscles arises from the anterior surfaces of the radius and ulna and the adjacent interosseous membrane. The **pronator quadratus** is a small square muscle, running transversely between the lowest quarters of the anterior surfaces of the radius and ulna. The **flexor pollicis longus,** arising from the anterior surface of the radius distal to the tuberosity and the oblique line as far as the attachment to the pronator quadratus, can be followed as a tendon deep to the flexor retinaculum as can the four tendons of the **flexor digitorum profundus,** arising from the anterior surface of the ulna. Remove the long extensor muscles from the lateral epicondyle and examine the **supinator muscle.** It is attached to the common extensor origin and to the lateral aspect of the upper end of the ulna just below the radial notch. The fibres run obliquely and laterally round the back of the upper end of the radius to be attached to the body above the oblique line. Passing backwards through the supinator muscle is the posterior interosseous branch of the radial nerve.

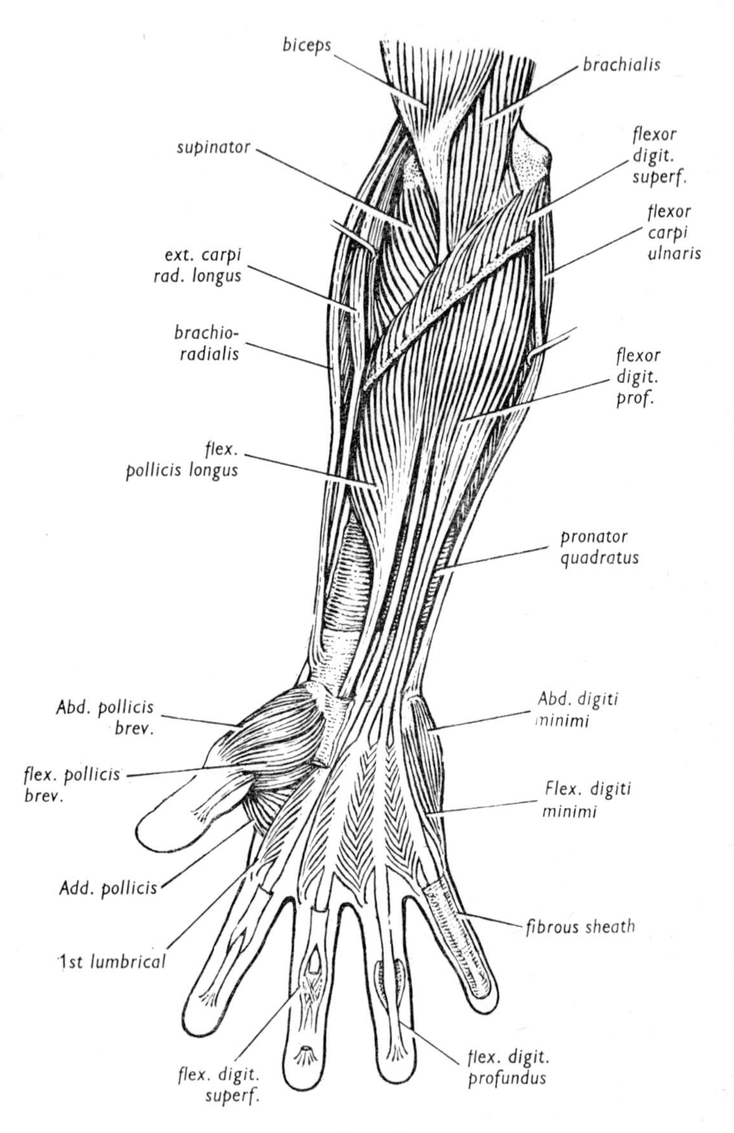

FIG. 20
The deep muscles of the front of the forearm and the superficial muscles of the hand.

STRUCTURAL DETAILS

The radius

The body of this bone below the radial tuberosity is slightly bowed laterally and has a sharp medial (interosseous) border to which the interosseous membrane is attached. The bone ends below as the styloid process which projects beyond the enlarged lower end of the bone. The anterior surface of the radius is crossed above by an obliquely running line passing downwards and laterally from the radial (bicipital) tuberosity and becoming the anterior border. The lateral surface has on its most prominent part of a roughened area for the insertion of the pronator teres. The posterior surface is smooth and gives attachment to muscles. The lower end of the bone is flattened in front and is partly subcutaneous (feel the radial pulse against the bone on your own wrist). The lateral aspect of the lower end is grooved and is continuous with the styloid process. The posterior surface is also grooved. The deepest groove is about the middle and is limited laterally by the **radial tubercle** which can be felt through the skin and frequently seen when the hand is strongly palmar flexed. On the medial side is the notch for articulation with the lower end of the ulna.

The ulna

Like the radius, the ulna has a body with a prominent interosseous border continuous above with the supinator crest behind the radial notch. The medial and posterior surfaces are subcutaneous distally. The posterior border is subcutaneous from the olecranon proximally to the styloid process distally. The lower end of the ulna is much smaller than that of the radius, is rounded and is covered by articular cartilage except for the medial aspect and the styloid process. Lateral to this process on the posterior aspect of the bone is a groove for the tendon of the extensor carpi ulnaris. The styloid process of the radius projects about 2 cm beyond that of the ulna.

The bones of the hand

The fundamental arrangement in mammals is for the radius and ulna to be fixed relative to each other and to articulate with the proximal row of three carpal bones—the scaphoid, lunate and triquetrum from the radial to the ulnar side. Distal to the lunate there is a central carpal, which in man is fused with the scaphoid. There is a row of five distal carpals, reduced to four in man by the fusion of the 4th and 5th. From the radial to the ulnar side, these are called the trapezium, trapezoid, capitate and hamate.

The **scaphoid** is the largest bone in the proximal row. Note that the **tubercle** which is on the distal palmar part of the bone, projects against the skin when the hand is dorsiflexed. It can be seen on the palmar aspect of the wrist, proximal to the ball of the thumb. The scaphoid is sometimes fractured. This results in pain on pressure in the " anatomical snuff box." The **lunate** and **triquetrum** complete the proximal row. The scaphoid and lunate articulate with the radius, and the triquetrum with the articular disc distal to the ulna. The **pisiform** is a sesamoid bone in the tendon of the flexor carpi ulnaris and articulates with the palmar surface of the triquetrum. In the distal row, the **capitate** in the middle is the largest bone of the carpus and its main proximal articulation is with the lunate. The **hamate** is placed medially, has a well marked hook on the palmar surface and distally articulates with two metacarpal bones. The **trapezium,** the most lateral carpal, has a well marked tubercle on the palmar surface and articulates by a saddle-shaped joint with the thumb metacarpal. The small **trapezoid** lies between the trapezium and the capitate.

There is considerable variation in the times of appearance of the centres of **ossification** in this region (Fig. 18). The distal epiphysis of the radius appears at about two years and that of the ulna at about five years. They both join the bodies about the twentieth year. Ossification of the carpus occurs between the first and the sixth years, and the pisiform ossifies later (about the twelfth year). The capitate is the first to ossify.

The muscles

The actions of the muscles of the upper arm and forearm will be considered in two parts. Those muscles acting on the elbow joint or the radio-ulnar joints are dealt with in the next chapter and those which act on the wrist and fingers are dealt with in the chapter on the hand (page 74).

The vessels and nerves

The **median nerve** enters the forearm between the humeral and ulnar heads of the pronator teres muscle and passes to the wrist. The nerve lies on the deep surface of the flexor digitorum superficialis and then on the lateral side of its tendons. Posteriorly is the flexor digitorum profundus and laterally at the wrist is the flexor carpi radialis tendon. When the palmaris longus is absent the nerve is subcutaneous at the wrist and is then more liable to injury. If the palmaris longus is present it is superficial and medial to the nerve. The main trunk enters the palm deep to the flexor retinaculum. Numerous muscular branches are given off between the elbow and wrist and supply the flexor muscles of the forearm except the flexor carpi ulnaris and the medial half of the flexor digitorum profundus. Cutaneous branches supply the palm of the hand.

The **ulnar nerve** passes behind the medial epicondyle and enters the forearm between the two heads of the flexor carpi ulnaris. In the forearm it is overlapped on the medial side by this muscle and lying deeply are the flexor digitorum profundus and pronator quadratus. The nerve, with the ulnar artery on its lateral side, lies medial to the flexor digitorum superficialis. Branches are given off to the flexor carpi ulnaris and the medial half of the flexor digitorum profundus. The main nerve trunk passes superfically into the palm and is covered by a slip of the flexor retinaculum. It lies on the lateral side of the pisiform bone. It gives off palmar and dorsal cutaneous branches.

The **radial nerve** has been traced down the forearm under cover of the brachioradialis. About 5 cm above the radial styloid process the nerve passes backwards and crosses the wrist to supply the skin on the back of the hand and lateral fingers. Through its posterior interosseous branch it supplies the muscles of the back of the forearm, except the brachioradialis and extensor carpi radialis longus which receive branches from the radial nerve above the lateral epicondyle.

The **ulnar artery** is one of the terminal branches of the brachial artery. It passes deep to the pronator teres and then with the median nerve deep to the flexor digitorum superficialis to reach the lateral side of the ulnar nerve deep to the flexor carpi ulnaris. It retains this position and passes with the nerve lateral to the pisiform bone into the palm.

The **radial artery** is the other terminal branch of the brachial

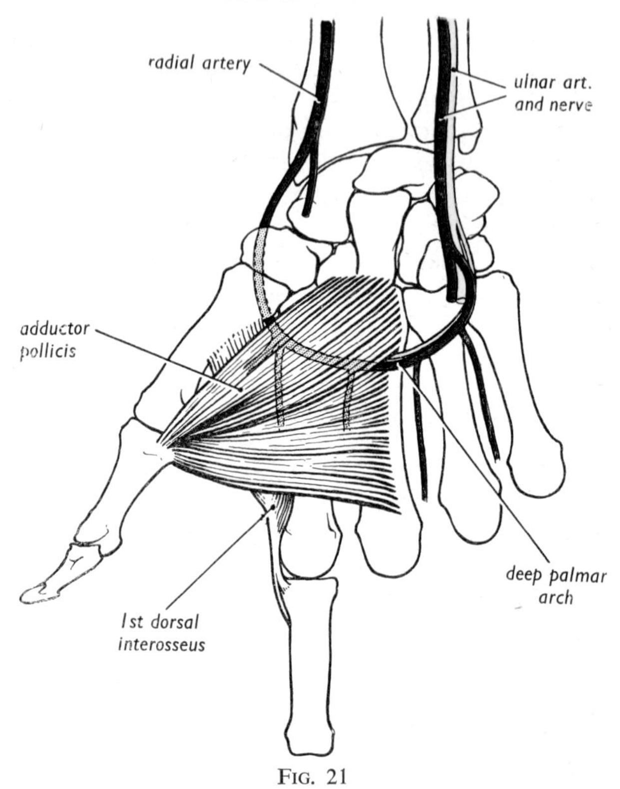

FIG. 21
The formation of the deep palmar arch from the radial artery and the deep branch of the ulnar artery.

artery. As it passes to the wrist, it lies deep to the brachioradialis on the muscles attached to the front of the radius. At the wrist it lies on the radius lateral to the tendon of the flexor carpi radialis where its pulsation can easily be felt (Fig. 21). The artery passes backwards round the lateral side of the wrist deep to the tendons of the abductor pollicis longus and extensor pollicis brevis and across the floor of the " anatomical snuff box." It then passes into the palm between the two heads of the first dorsal interosseus muscle in the first intermetacarpal space. Both the radial and ulnar arteries give off many muscular, articular and cutaneous branches, as well at nutrient arteries to the radius and ulna before entering the palm.

CHAPTER 9

THE ELBOW AND RADIO-ULNAR JOINTS

INTRODUCTION

THE elbow is a compound joint permitting flexion and extension at the humeroradial and humero-ulnar joints and also supination and pronation at the proximal radio-ulnar and the humeroradial joints.

Revise the muscles of the upper arm that produce flexion and extension at the elbow joint (pages 51 and 52).

DISSECTION

Remove all muscle tissue around the capsule of the elbow joint. Note that the capsule is thin anteriorly and posteriorly and thickened at the sides to form the medial and lateral ligaments. It is separated from the triceps by a bursa. Proximally, in front, the capsule is attached to the edge of the articular cartilage on the humerus but posteriorly it crosses the floor of the olecranon fossa. Distally, the capsule is attached to the medial edges of the olecranon and coronoid process of the ulna, and, lateral to the tip of the coronoid process, it is attached to the **annular ligament.** This is a strong fibrous band that encircles the head of the radius, holds it in contact with the ulna and is attached to the latter bone in front of and behind the radial notch. The capsule of the elbow joint is attached along the lateral edge of the olecranon above the attachment of the annular ligament.

The **ulnar** and **radial collateral ligaments** are thickenings of the capsule and are attached to the humerus on the rough areas between the articular edge and the epicondyles. The fibres of both radiate downwards, the lateral to the annular ligament and the medial to the medial side of the coronoid process and olecranon. The ulnar nerve is closely related to the medial ligament.

Open the capsule by transverse incisions in front and behind at the level of the distal surface of the humerus. Note that the capsule is lined with synovial membrane which extends to the edge of the articular cartilage. Both in front and behind there are pads of fat between the capsule and synovial membrane opposite the olecranon and coronoid fossae of the humerus. Remove as much capsule as possible from the annular ligament and observe the movements that are possible at the proximal radio-ulnar joint. The radial tuberosity passes backwards in pronation. Pull on the biceps tendon when the forearm is pronated and note the effect on the radius.

Separate completely the humerus from the radius and ulna and examine the articular surfaces of the three bones. The medial pulley-shaped articular surface (trochlea) for the ulna is separated by a marked ridge from the round surface (capitulum) for the radius.

Incise the anterior attachment of the annular ligament and examine the proximal radio-ulnar joint. This is a pivot joint, allowing rotation round a longitudinal axis passing through the head of the radius and the lower end of the ulna.

Carefully separate the radius and ulna by cutting the interosseous membrane and ligaments. Note that the direction of the fibres of the membrane is downwards and medial from the radius to the ulna. The distal radio-ulnar joint will be opened but do not cut the cartilaginous disc which unites the lower end of the radius to the ulna and separates the ulna from the wrist joint.

The **distal radio-ulnar joint** is a pivot joint allowing the movements necessary for pronation and supination of the forearm. The capsule encloses the joint cavity which extends upwards for a short distance between the radius and ulna. Note that the ulna has a lateral facet for articulation with the radius and a distal facet for articulation with the **articular disc.** This disc is made of fibrocartilage and helps to hold the radius and ulna together. It is triangular in shape. The apex is attached to a pit on the lateral side of the styloid process of the ulna and the base to the lower border of the ulnar notch of the radius. The cavity of the distal radio-ulnar joint does not usually communicate with that of the wrist joint.

FUNCTIONAL ASPECTS

The movements at the elbow and radio-ulnar joints

Movements occur round two principal axes: (1) transverse,

allowing flexion and extension of the forearm at the humeroradial and humero-ulnar joints and (2) longitudinal, allowing pronation and supination of the forearm at the radio-ulnar and radiohumeral joints. Flexion is mainly produced by the brachialis and the biceps, and extension by the triceps. The muscles of the forearm that are attached to the epicondyles of the humerus may help these movements. Pronation is produced by the pronator teres and the pronator quadratus. The axis of rotation can pass through the head of the radius and the lower end of the ulna although the axis may vary and can pass through the tip of any of the medial four digits. The head of the radius rotates on the capitulum of the humerus as well as on the ulna, which apparently moves slightly backwards and laterally. Supination is brought about by the biceps and the supinator. The force that can be exerted in both pronation and supination is greatest when the elbow is flexed to about 90°. Usually supination is a more powerful movement than pronation.

CHAPTER 10

THE PALM OF THE HAND

INTRODUCTION

THE palm of the hand and the palmar surfaces of the fingers form a very sensitive area of the skin surface. It is richly supplied with nerves and blood vessels. Not only is the hand an efficient tactile organ but it is capable of careful, accurate and fine movements due to the short muscles, as well as strong and sustained grasping movements due to the long flexors. The nerves supplying the long flexors and the small muscles come from the median and ulnar nerves. The forearm muscles are mostly supplied from the 7th and 8th cervical spinal nerves and the hand muscles from the 1st thoracic spinal nerve.

The skin has been removed from the palm of the hand. Note that it varies in thickness, is ridged, devoid of hair and in some places firmly attached to the deeper tissues forming flexure lines. The superficial fascia contains fat but there is more fibrous tissue than usual. The deep fascia is much thickened to form the palmar aponeurosis, continuous proximally with the flexor retinaculum and distally along the fingers as a fibrous tunnel in which the tendons of the flexors lie. The muscle spaces for the thumb and the little finger are separated from the tendon space in the centre by septa of fibrous tissue running from the palmar aponeurosis to the 1st and 5th metacarpal bones respectively.

On the articulated skeleton identify the tubercle of the scaphoid, the pisiform, the tubercle of the trapezium and the hook of the hamate in the carpus. The first two are also easily palpated in the cadaver and on yourself. Note the positions of the carpometacarpal, metacarpophalangeal and interphalangeal joints. The dorsal aspects of the metacarpal and phalangeal bones can be easily palpated.

DISSECTION

Remove the fascia covering the thenar and hypothenar muscles and the palmar aponeurosis from the middle of the palm by cutting its attachment to the hamate and trapezium. Take care to preserve the ulnar nerve and vessels. Turn the aponeurosis towards the fingers, exposing the long tendons on which lie the superficial palmar arterial arch and branches of the ulnar and median nerves. Dissect the thenar and hypothenar eminences, each containing three muscles—an abductor near the margin of the hand, a flexor nearer the midline, and an opponens deep to the other two. The thumb muscles are the abductor pollicis brevis, flexor pollicis brevis and opponens pollicis. The little finger muscles are the abductor digiti minimi, the flexor digiti minimi and the opponens digiti minimi.

Note that owing to the rotation of the thumb medially, the movements of flexion, extension, abduction and adduction of this digit occur in planes at right angles to the planes in which these movements occur at other joints in the limb.

Define the attachments of the flexor retinaculum again. The **abductors** come from a proximal bone (scaphoid or pisiform) and the retinaculum. The thenar abductor is attached distally to the radial side of the proximal phalanx of the thumb, and the hypothenar abductor to the ulnar side of the base of the proximal phalanx of the little finger.

The **flexor** and **opponens** come from the distal retinacular attachment (the tubercle of the trapezium or the hook of the hamate). Each flexor is inserted with the abductor, and the opponens is inserted along the body of the 1st or 5th metacarpal bone. The fibres of the flexor and abductor are longitudinal and those of the opponens are very oblique. These muscles can best be identified by defining their attachments and then cleaning the intermediate muscle belly. To expose the opponens it is necessary to cut transversely through the abductor. The adductor pollicis cannot be fully seen since it lies deep to the long tendons.

Deep to the palmar aponeurosis find (1) the superficial palmar arterial arch a continuation of the ulnar artery, (2) the branches of the median nerve to the fingers (cutaneous) and to the thenar muscles and (3) the branches of the ulnar nerve to the fingers (cutaneous) and to the hypothenar muscles. Clean the digital branches of the nerves and arteries and note that the ulnar nerve is usually distributed to the medial one and a half digits and the median nerve to the lateral three and a half digits. Compare the arrangement in various dissected hands. Find a lumbrical muscle on the lateral side of each flexor digitorum profundus tendon and note the small branch of the digital nerve which supplies it. The medial two lumbricals are usually supplied by the ulnar nerve and the lateral two by the median nerve.

Cut the median nerve at the wrist and turn it distally. Cut the **ulnar artery** as it turns laterally to form the superficial arch and remove small branches so that the arch is freed. Note that the ulnar artery gives off a deep branch accompanying the deep branch of the nerve round the medial side of the hook of the hamate into the deeper part of the palm, and that the superficial arch is completed laterally by anastomosis with a superficial branch of the radial artery. Examine the long flexor tendons, which are separated by loose fascia from the metacarpal bones, the palmar interossei and the adductor pollicis. Cut the long flexor tendons above the wrist and pull the cut ends towards the fingers, thus exposing fully the deepest structures in the palm. A thin fibrous septum passes from the deep surface of the palmar aponeurosis, medial to the long flexor tendons of the index finger, to the 3rd metacarpal bone. This divides the deep fascial space into a lateral **thenar space** containing the flexor pollicis longus tendon and the flexor indicis tendons with the associated lumbrical muscle and a medial **midpalmar space** containing the other long flexor tendons and their associated lumbrical muscles.

Clean the lumbrical muscles, and note their attachments to the profundus tendons. Remove any skin and the fibrous tunnel from the palmar surface of one or two fingers and the thumb (Fig. 20). Trace the tendons of the lumbrical muscles round the radial side of the metacarpophalangeal joints to their attachment to the extensor expansions. The muscles act as flexors at the metacarpophalangeal joints and extensors at the interphalangeal joints. Examine the arrangement of the long flexor tendons inside the flexor fibrous tunnel. Each **superficialis** tendon divides and is attached to the sides of a middle phalanx. Each **profundus** tendon passes through a superficialis tendon and is attached to the base of a terminal (distal) phalanx. The tendon of the **flexor pollicis longus** can be followed to its insertion on the base of the terminal phalanx of the thumb. Examine the tendons carefully and note the extent of the synovial sheaths in the palm and fingers.

Clear away the loose areolar tissue from the surface of the metacarpal bones and the adductor pollicis and define the deep branch of the ulnar nerve and the deep palmar arterial arch formed by the radial artery anastomosing with the deep branch of the ulnar artery (Fig. 21). Clean the transverse head of the adductor pollicis, separate it from the middle metacarpal and turn it towards the thumb thus exposing the lateral interosseous spaces. The **adductor pollicis** lying partly in the web of the thumb is inserted on to the ulnar side of the base of the proximal phalanx of the thumb and arises by two heads, an oblique from the bases of the 2nd and 3rd metacarpal and adjacent carpal bones and a transverse from the anterior surface of the shaft of the 3rd metacarpal bone. Clean the palmar interrosseus muscles and trace the tendons to their attachments to the proximal phalanges and the extensor expansions. It will be necessary to cut the **deep transverse ligaments** of the palm, which bind together the adjacent metacarpal heads. The lumbrical tendons pass in front of these ligaments and the interosseus tendons behind them.

The four **palmar interossei** lie between the metacarpal bones. Each is attached to one metacarpal body (1, 2, 4 and 5) and acts as an adductor of the corresponding finger (Fig. 19). The tendons are attached to the ulnar side of the base of the proximal phalanx of the thumb and index fingers and to the radial side of the base of the proximal phalanx of the ring and little fingers. Fibres are also attached to the extensor expansions on the back of the fingers.

The **ends of the fingers** are highly specialised in a number of ways. Remove the skin from the pad of a finger. Cut into the pad and note that it consists of lobules of fat separated by fibrous septa passing between the skin and the periosteum of the phalanx. Make a longitudinal incision from the distal interphalangeal joint through the nail bed to the tip of the finger. Note that the root and sides of the nail are buried by neighbouring skin and that the nail can be more easily detached from the nail bed in its proximal part than distally.

STRUCTURAL DETAILS

The metacarpal and phalangeal bones

Articulating with the distal carpal bones and lying in the palm of the hand are the five metacarpal bones. Distal to these are the five digits, the first containing two and the others three phalanges. The metacarpal bones and the phalanges each have a body, enlarged proximally and distally to form a **base** and **head** respectively. The articular surface of the head continues on to the palmar surface of the bone. Each terminal phalanx has an irregular head that is non-articular and flattened. The primary centres of **ossification** for the bodies of the metacarpals and phalanges appear at about the end of the second month of fetal life (Fig. 18). The secondary centres for the heads of the 2nd, 3rd, 4th and 5th and the base of the 1st metacarpals and the bases of the phalanges appear about the third year and join the bodies at about the eighteenth year.

The nerves and vessels

The **median nerve** after entering the palm deep to the retinaculum divides into three or four digital branches to the two sides of the thumb, index and middle fingers and the lateral half of the ring finger, and a branch to supply the three thenar muscles. The digital branches as they pass towards the first and second webs give off twigs to the first two lumbricals. The digital nerves extend along the sides of the fingers and supply not only the palmar surface but also the dorsal surface of the terminal phalanges including the nail beds.

The **ulnar nerve** enters the palm superficially over the flexor retinaculum but covered by a process of deep fascia (Fig. 21). Near the hook of the hamate it divides into a superficial and a deep branch. The superficial branch crosses the hook of the hamate and gives off digital branches to both sides of the little finger and the medial side of the ring finger and muscular twigs to the medial two lumbricals. The digital branches also supply the dorsal surface of the terminal phalanges. The deep branch passes medial to the hook of the hamate and then laterally between the abductor and the flexor digiti minimi, supplying the muscles of the hypothenar eminence. It reaches the palm deep to the long tendons and lies proximal to the deep arterial arch. The terminal branches pass between the oblique and transverse heads of the adductor pollicis. The deep branch supplies them and all the interossei.

The **radial artery** enters the palm between the two heads of the first dorsal interosseus, deep to the adductor pollicis. It emerges between the two heads of adductor pollicis, runs

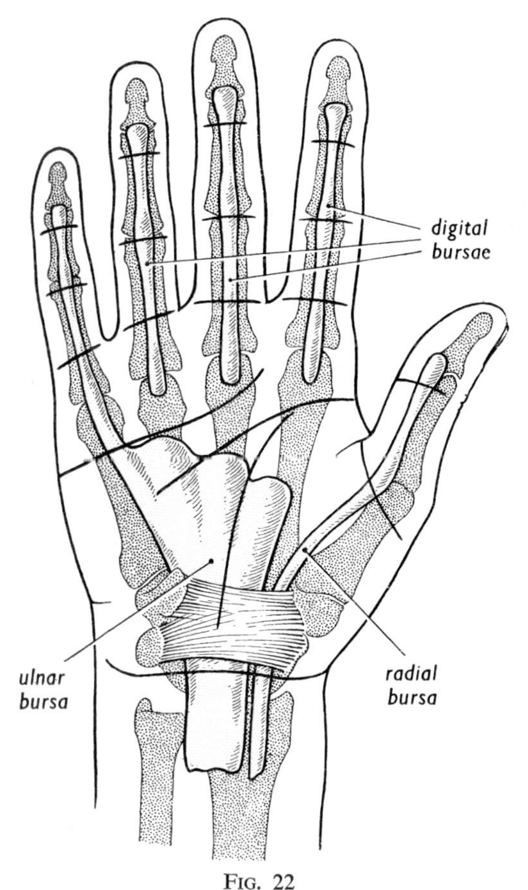

FIG. 22
The extent of the synovial sheaths in the hand and fingers. Note also the relationships of the skin flexure lines to the underlying joints.

across the palmar surface of the metacarpal bones and interosseus muscles and anastomoses with the deep branch of the ulnar artery (which accompanies the deep branch of the ulnar nerve) thus completing the deep palmar arch. From this arch digital and muscular branches are given off. Arising from the radial artery as it lies deep to the adductor pollicis are the main branches to the thumb and to the index finger (Fig. 21).

The synovial sheaths (Fig. 22)

In the hand and at the wrist the long tendons are surrounded by synovial sheaths, except in the palm in the region of the attachment of the lumbrical muscles. The tendon of the flexor pollicis longus is surrounded by a synovial sheath **(radial bursa)** extending from 3 cm above the wrist to its attachment to the distal phalanx. The common flexor sheath **(ulnar bursa)** for the flexor digitorum superficialis and flexor digitorum profundus commences at about the same level above the wrist and extends to the mid-palm at the level of the proximal transverse skin crease. The sheath is continued along the tendons of the little finger to the base of its terminal phalanx. The tendons of the index, middle and ring fingers have no synovial sheath for some distance but the sheath is present again from just beyond the distal transverse skin crease to the bases of the terminal phalanges.

The sheaths not only reduce friction, but have a protective action on the long, relatively avascular tendons. One layer of the synovial membrane is closely applied to the tendons and the other lines the fibro-osseous space in which the tendons lie. The two layers are continuous at the ends of the sheath and developmentally there is a mesentery from the phalanges to the tendon and this transmits blood vessels. This mesentery disappears but remnants of it exist as several vincula, which attach the long tendons to the bones near the interphalangeal joints (Fig. 19).

CHAPTER 11

THE JOINTS OF THE WRIST AND HAND

INTRODUCTION

MOVEMENTS at these joints are produced by the long muscles of the forearm and the short muscles of the palm. The former are the more active in such movements as gripping and the latter in writing or dissecting.

Revise the osteology of the wrist and hand (pages 67 and 72).

DISSECTION

Remove any remnants of muscle attached to the hand. Determine the movements that are possible at the wrist and try to analyse the joints participating. The tubercle of the scaphoid, for example, moves backwards on flexion at the wrist. Note the relative positions of the radial and ulnar styloid processes and open the joint cavity posteriorly between the radius and ulna proximally and the carpal bones distally. On the distal surface of the radius are the facets for the scaphoid and lunate, and medial to these facets is the **articular disc** which articulates with the triquetral and separates the distal end of the ulna from the wrist joint. Cut through the ulnar attachment of the articular disc and completely separate the radius from the ulna.

On the palmar surface of the wrist note the ligaments radiating from the capitate bone to almost all the other bones of the wrist. The ligaments on this surface are much stronger than those on the dorsal surface of the joints.

Open up the **intercarpal joint** space dorsally between the proximal and distal rows of carpal bones. This joint cavity extends across the wrist but does not usually communicate with either the radiocarpal or carpometacarpal joints. The capitate bone is the largest in the distal row and the upper end of the bone fits into the hollow formed by the scaphoid and lunate.

Open up the **carpometacarpal joints** dorsally and notice the shape of the joint between the trapezium and the first metacarpal bone. The saddle-shape permits movements in several planes. The 4th and 5th carpometacarpal joints permit a limited degree of passive movement but there is no movement at the carpometacarpal joints of the index and middle fingers.

The carpal bones are bound together by dorsal and palmar **carpal ligaments,** and the sides are strengthened by **medial** and **lateral ligaments; interosseous ligaments** are placed deeply between the bones.

Open up the dorsal aspect of the **metacarpophalangeal** and **interphalangeal joints** of a finger. The larger articular surface is on the proximal bone and extends on to the palmar surface for some distance. On your own fingers work out the position of the line of the joints relative to the skin creases. The capsule of these joints dorsally is formed by the expansion of the extensor tendons. On each side a **collateral ligament** runs distally and forwards, and on the palmar surface there is a much thickened fibrous pad **(palmar ligament),** more firmly attached distally than proximally. When the finger is flexed this pad moves proximally on the palmar articular surface of the head of the proximal bone (metacarpal or phalanx).

FUNCTIONAL ASPECTS

The movements at the wrist consist of flexion (palmar flexion), extension (dorsiflexion), abduction (radial deviation), adduction (ulnar deviation) and circumduction. These movements take place at the radiocarpal and intercarpal joints. Taken as a whole this joint is ellipsoid in nature. The intercarpal joints allow both flexion and extension and some side to side movements. The carpometacarpal joints, except that of the thumb, permit little movement, but the little finger can be opposed to a slight extent. The carpometacarpal joint of the thumb is saddle-shaped and permits a considerable amount of flexion, extension, adduction, abduction and rotation. Owing to the medial rotation of the thumb at the carpometacarpal joint the terms used to describe the movements of the thumb do not have their usual meaning. Thus the thumb is said to be abducted when it is moved away from the palm in a plane at right angles to the palm and flexed when it is bent medially in the plane of the palm. Similarly adduction is pressure of the side of the thumb against the index finger and extension is movement away from the hand in the plane of the palm. Opposition is the movement across the palm so that the tip of the thumb meets the tips of the other fingers and is a combination of abduction then flexion, adduction, and medial rotation.

The metacarpophalangeal and the interphalangeal joints of the thumb and fingers permit flexion and extension. Abduction and adduction also occur at the metacarpophalangeal joints, so allowing spreading of the fingers (Fig. 19 B and C). Abduction and adduction are very limited when the metacarpophalangeal joints are flexed.

The names of many of the muscles of the forearm and hand indicate which muscles produce the movements described, *e.g.* the flexors carpi radialis and ulnaris flex the hand at the wrist. At this joint adduction (ulnar deviation) is produced by the flexor carpi ulnaris and extensor carpi ulnaris acting together and abduction (radial deviation) by the flexor carpi radialis and extensors carpi radialis longus and brevis. The flexor digitorum profundus produces flexion at the distal interphalangeal joints and then flexion at the more proximal joints, and the flexor digitorum superficialis acts primarily on the proximal interphalangeal joints. Extension at the interphalangeal joints is produced by the interossei and lumbricals in association with the extensor digitorum. Flexion at the metacarpophalangeal joints is produced by the lumbricals and extension by the extensors digitorum, indicis and digiti minimi.

The most important of the coarser movements are those involved in grasping. The long flexor muscles bring this about and the wrist is immobilised by fixators—the flexors and extensors carpi ulnaris and radialis. The most important fine movements occur between the thumb and index finger. In writing, the thumb is opposed so that the pad of the thumb is directed towards the other fingers. Pressure by the thumb on the fingers is exerted by the adductor pollicis. In writing, a down-stroke is produced by flexion of the interphalangeal joints (the long flexors) and the up-stroke by extension of the interphalangeal joints (the lumbricals and interossei). Carriage across the paper is produced by extension at the elbow joint and by lateral rotation at the shoulder joint.

A precision grip involves holding an object between the tip of the thumb and the tip of one or more fingers. A power grip involves holding an object firmly in the palm of the hand by means of the strongly flexed fingers with the thumb overlapping the object. The shape and weight of the object also play a role in how it is held by the hand.

CHAPTER 12

THE CUTANEOUS NERVE SUPPLY AND THE LYMPH DRAINAGE

THE CUTANEOUS NERVE SUPPLY

As the limb buds grow out from the body of the embryo the skin is stretched over the developing muscle and bone. The upper limb appears opposite the 4th cervical to the 2nd thoracic spinal cord segments. Fibres from the 7th cervical nerve pass to the extremity of the limb and are distributed to a strip of skin down the centre of the front of the forearm and hand from the middle of the forearm above to the tip of the middle finger below (Fig. 23). On the back the strip extends from the elbow region above to the tip of the middle finger below.

The lateral (pre-axial) border of the upper limb is supplied by the 4th, 5th and 6th cervical spinal nerves (the 4th supplies the region of the shoulder, the 5th the upper arm and the 6th the forearm and hand). The medial border (postaxial) is supplied by the 8th cervical and 1st and 2nd thoracic spinal nerves (the 8th cervical supplies the hand and distal forearm, the 1st thoracic the region above and below the elbow and the 2nd thoracic the rest of the upper arm). The area supplied by each spinal nerve extends from the lateral or medial border on to the anterior and posterior surfaces of the limb. It should be noted that the area supplied by one spinal nerve is also supplied to some extent by the neighbouring spinal nerves. The areas supplied by the 3rd to the 6th cervical nerves lie adjacent to those supplied by the 1st, 2nd and 3rd thoracic nerves. In front the line separating these two groups of areas is the **ventral (anterior) axial line.** It extends from a point lateral to the manubriosternal joint to a point near the middle of the forearm. The **dorsal (posterior) axial line** extends from the upper scapular region down the middle of the back of the upper arm to the elbow joint. Both the dorsal and ventral axial lines end at the point where the distribution of the 7th cervical spinal nerve reaches the surface.

The larger cutaneous nerves are shown in Figures 24A and 24B.

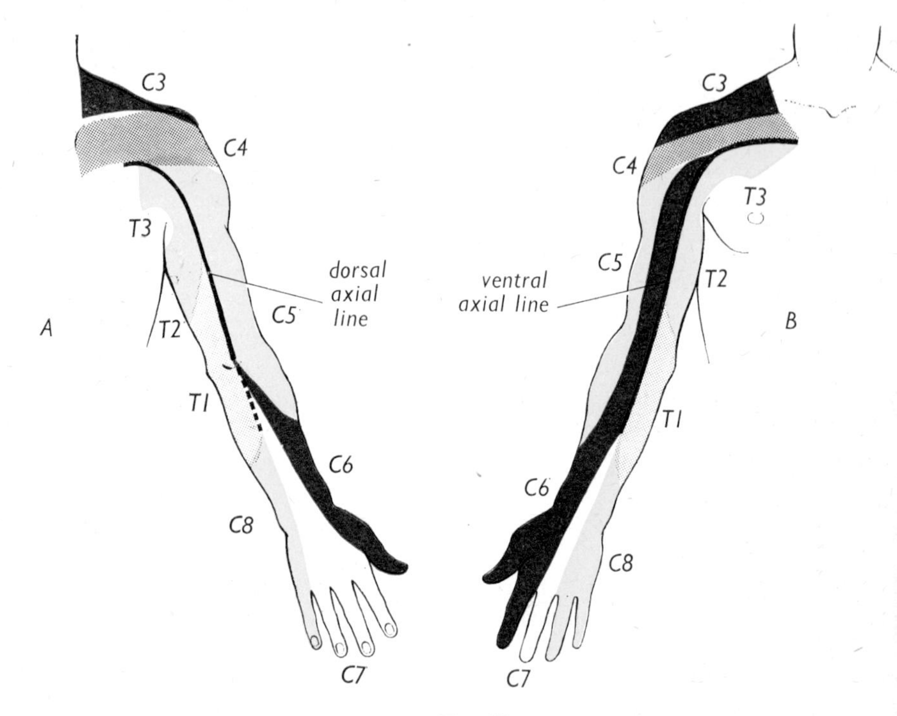

FIG. 23

The pattern of the dermatomes of the upper limb. A, the dorsal, B, the ventral aspects. The heavy lines indicate the positions of the dorsal and ventral axial lines.

THE LYMPH DRAINAGE

As in most parts of the body the lymph vessels of the upper limb are arranged in two groups, (1) superficial—draining the skin and subcutaneous tissues, and (2) deep—draining the muscles, etc. Both groups of vessels drain into the axillary nodes.

The **superficial lymph vessels** run as long, interconnected channels under the skin. They receive tributaries from the hair follicles, glands, etc., of the skin. On the medial side of the forearm the vessels run with the basilic vein and in the region of the medial epicondyle they are connected to a **cubital lymph node.** The efferents from this node pass with the basilic vein to join the deeper lymph channels that run with the brachial veins or

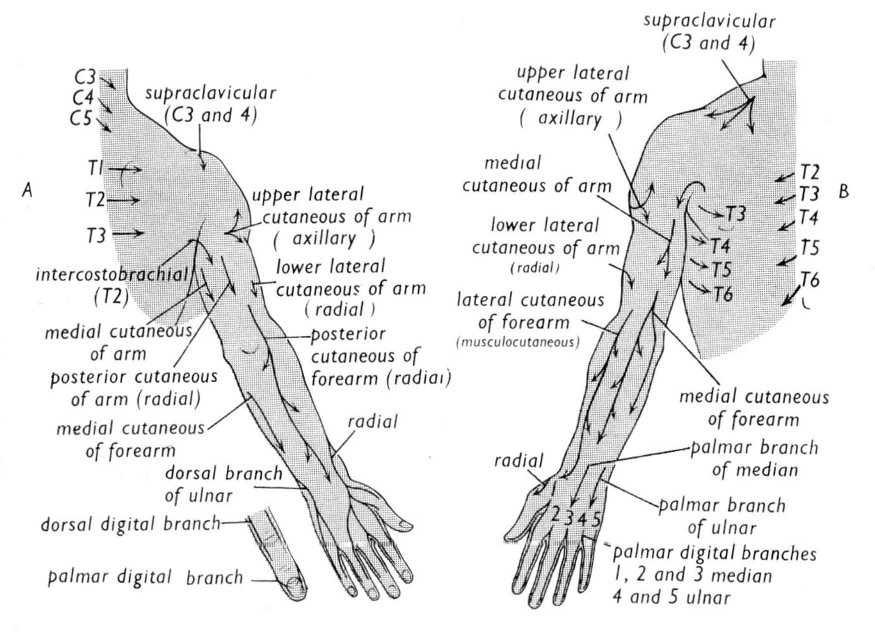

FIG. 24
The arrangement of the cutaneous nerves on A, the dorsal and B, the ventral aspects of the upper limb.

pass superficially to the axilla. On the lateral side the lymph vessels run with the cephalic vein as far as the infraclavicular fossa before joining a lymph node.

The **deeper vessels** run with the deep veins and drain the muscles and joints. They drain into the lateral (humeral) group of the axillary nodes and their efferents pass to the central and apical axillary nodes.

The axillary nodes (Fig. 25)

These nodes are arranged in groups: (1) deep to the lower edge of the pectoralis major on the lateral aspect of the chest wall **(pectoral)**; (2) on the lateral axillary wall **(lateral)**; (3) on the posterior axillary wall **(subscapular)**; (4) in the fat of the axilla

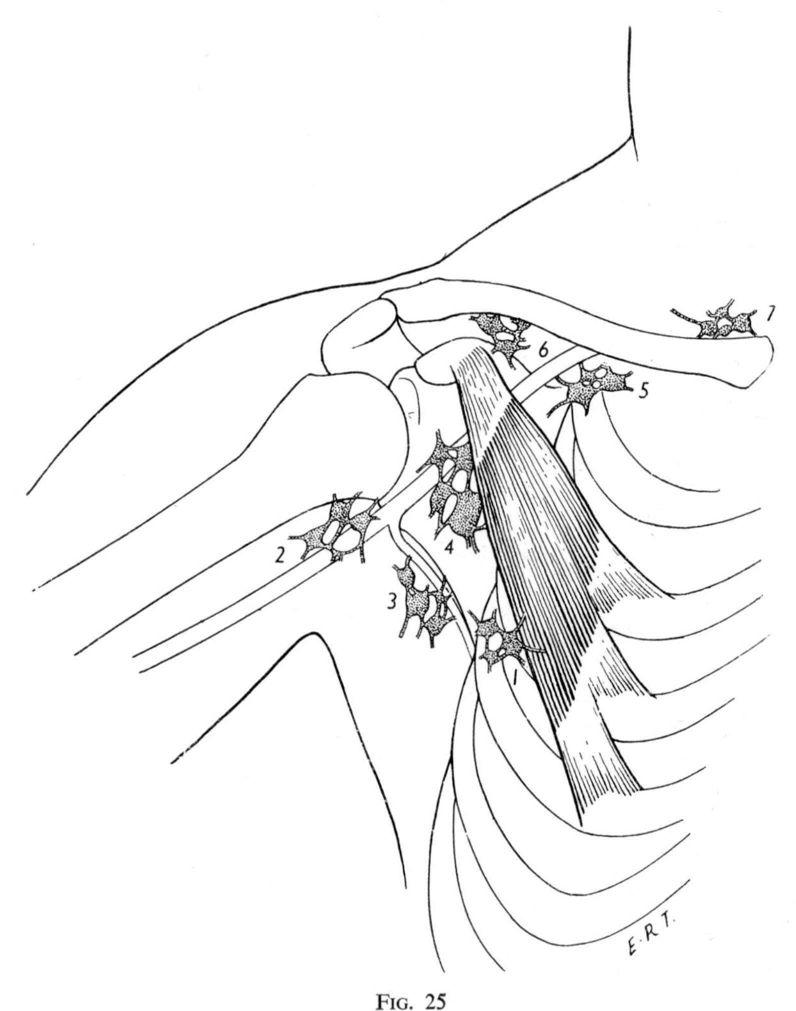

FIG. 25
The arrangement of the axillary lymph nodes. 1. pectoral; 2. lateral; 3. subscapular; 4. central; 5. apical; 6. infraclavicular; 7. supraclavicular or lower deep cervical. The axillary vein is indicated.

between the other three **(central)**; (5) at the apex of the axilla **(apical),** along the axillary vein, behind and then above the clavicle and continuous with the most inferior of the deep cervical nodes of the neck. The lateral group receives almost all the lymph from the upper limb, the pectoral groups receives lymph from the mammary gland as well as from the lateral and anterior chest wall, and the subscapular, from the back of the trunk. The efferents drain into the central and apical groups. The efferents from the apical group form the **subclavian lymph trunk,** which joins with the **jugular trunk** from the head and neck and forms a larger trunk. On the right, this trunk joins the right brachiocephalic vein, and on the left the thoracic duct, before it enters the left brachiocephalic vein. (6) Along the cephalic vein there may be a few nodes **(infra-clavicular group)** whose afferents come from the upper limb and lateral part of the breast and whose efferents drain into the apical group.

PRACTICAL CLASS 1

This is a guide to what should be known about individual bones and the articulated skeleton.

THE OSTEOLOGY OF THE SCAPULA, CLAVICLE AND SHOULDER JOINT

Requirements: Articulated skeleton; separate clavicle, scapula and humerus.

1. The scapula

Place the scapula as it would lie in the upright body.

(*a*) Identify

1. the supraspinous, infraspinous and subscapular fossae,
2. the superior, medial and lateral borders,
3. the superior, inferior and lateral angles,
4. the spine,
5. the acromion,
6. the clavicular facet,
7. the coracoid process,
8. the neck,
9. the glenoid cavity,
10. the suprascapular notch.

(*b*) Show the approximate region of attachment of

1. the supraspinatus,
2. the infraspinatus,
3. the subscapularis,
4. the rhomboids,
5. the levator scapulae,
6. the serratus anterior,
7. the teres major,
8. the teres minor,
9. the long head of triceps,
10. the long and short heads of biceps brachii,
11. the coracobrachialis,
12. the pectoralis minor,
13. the omohyoid,
14. the trapezius,
15. the deltoid.

(*c*) Outline the attachment of

1. the capsule of the shoulder joint,
2. the glenoid labrum,
3. the coracoclavicular ligament,
4. the coraco-acromial ligament.

2. **The clavicle**

Place the clavicle as it would lie in the upright body.

(*a*) Identify the facet for
1. the acromion,
2. the sternum and 1st costal cartilage.

(*b*) Show the region of attachment of
1. the trapezius,
2. the sternocleidomastoid,
3. the pectoralis major,
4. the deltoid.

(*c*) Show the attachments of
1. the coracoclavicular ligament,
2. the costoclavicular ligament.

(*d*) What movements take place at the sternoclavicular and acromioclavicular joints?

(*e*) What stresses does the coracoclavicular ligament bear?

3. **The humerus**

Place the humerus as it would lie in the upright body.

(*a*) Identify
1. the head,
2. the anatomical neck,
3. the surgical neck,
4. the greater tuberosity,
5. the lesser tuberosity,
6. the intertubercular groove,
7. the body,
8. the deltoid tuberosity,
9. the radial groove.

(*b*) Show the approximate region of attachment of
1. the supraspinatus,
2. the infraspinatus,
3. the teres minor,
4. the subscapularis,
5. the pectoralis major,
6. the latissimus dorsi,
7. the teres major,
8. the coracobrachialis,
9. the brachialis,
10. the medial and lateral heads of the triceps,
11. the deltoid.

(*c*) Outline the attachment of the capsule of the shoulder joint.

PRACTICAL CLASS 2

THE ANATOMY OF THE SHOULDER REGION IN THE LIVING SUBJECT

Requirements: skin pencils, tape measures, subjects.

1. Mark the outline of the clavicle.
2. Is the lateral end of the clavicle higher than the medial end?
3. Mark the position of the sternoclavicular and the acromio-clavicular joints. Note the relation of the clavicle to the sternum and to the acromion.
4. Where is the infraclavicular fossa? What structures lie in this fossa?
5. Show the muscles that lie lateral and medial to this fossa.
6. Palpate the coracoid process and the acromion and mark their positions.
7. Show the muscle that forms the anterior axillary fold.
8. Show the muscles that form the posterior axillary fold.
9. Show the muscle on the medial wall of the axilla.
10. Mark the outline of the medial border of the scapula with the arm hanging at the side.
11. Mark the superior and the inferior angles and the spine of the scapula.
12. What vertebral spinous process is on a level (*a*) with the superior and (*b*) with the inferior scapular angle; (*c*) with the spine of the scapula?
13. Can the lateral border of the scapula be felt?

PRACTICAL CLASS 3

THE ANATOMY OF THE SHOULDER REGION IN THE LIVING SUBJECT (cont.)

1. Can the scapula move up and down without rotation? If so, which muscles raise and lower it in this way?

2. Show the muscles that rotate the glenoid cavity upwards.

3. At what angle of abduction of the arm does the scapula begin to rotate?

4. Outline the deltoid muscle and show the position of its attachments.

5. Show the muscles that produce at the shoulder joint
   1. flexion,
   2. extension,
   3. abduction,
   4. adduction,
   5. medial rotation,
   6. lateral rotation.

6. Show the actions of the following muscles
   1. the trapezius (the whole muscle, upper and lower fibres),
   2. the levator scapulae,
   3. the rhomboids,
   4. the latissimus dorsi,
   5. the serratus anterior,
   6. the deltoid (the whole muscle, anterior and posterior fibres),
   7. the supraspinatus,
   8. the infraspinatus,
   9. the teres major,
   10. the pectoralis major (the clavicular and sternocostal heads),
   11. the subscapularis.

7. Mark the position of
   1. the greater tuberosity of the humerus,
   2. the lesser tuberosity of the humerus,
   3. the axillary artery,
   4. the brachial plexus.

   With the arm slightly abducted and slightly laterally

rotated, draw a line from the midclavicular point to the mid-point of the cubital fossa. This line marks the position of the axillary and brachial arteries.

8. Show how the head and body of the humerus may be palpated.

PRACTICAL CLASS 4

THE OSTEOLOGY OF THE ELBOW JOINT AND FOREARM

Requirements: Articulated skeleton; separate humerus, radius and ulna; skin pencils and subjects.

1. **The humerus**

Place the humerus as it would lie in the upright body.

(*a*) Identify
1. the epicondyles,
2. the supracondylar ridges,
3. the trochlea,
4. the capitulum,
5. the olecranon fossa,
6. the coronoid fossa,
7. the radial fossa,
8. the body.

(*b*) Show the approximate area of attachment of
1. the common flexor origin,
2. the common extensor origin,
3. the brachialis,
4. the pronator teres,
5. the brachioradialis.
6. the extensor carpi radialis longus.

2. **The radius**

Place the radius as it would lie in the upright body.

(*a*) Identify
1. the interosseous border,
2. the head,
3. the neck,
4. the tuberosity,
5. the oblique line,
6. the styloid process,
7. the dorsal tubercle,
8. the groove for the extensor pollicis longus,
9. the ulnar notch.

(*b*) Show the approximate area of the attachment of

1. the supinator,
2. the biceps brachii,
3. the flexor digitorum superficialis.
4. the flexor pollicis longus,
5. the pronator quadratus,
6. the abductor pollicis longus,
7. the extensor pollicis brevis,
8. the pronator teres,
9. the brachioradialis.

3. **The ulna**

Place the ulna as it would lie in the upright body.

(*a*) Identify

1. the interosseous border,
2. the olecranon,
3. the trochlear notch,
4. the coronoid process,
5. the radial notch,
6. the supinator crest,
7. the head,
8. the styloid process,
9. the groove for the extensor carpi ulnaris.

(*b*) Show the approximate areas of attachment of

1. the triceps,
2. the brachialis,
3. the pronator teres,
4. the supinator,
5. the flexor digitorum profundus,
6. the pronator quadratus,
7. the flexor digitorum superficialis,
8. the abductor pollicis longus,
9. the extensor pollicis longus,
10. the extensor indicis.

4. Articulate the humerus, radius and ulna and show the attachments of

1. the capsule of the elbow joint,
2. the annular ligament,
3. the ulnar and radial collateral ligaments,
4. the interosseous membrane.

THE ANATOMY OF THE ELBOW AND FOREARM IN THE LIVING SUBJECT

1. Mark the position of the medial and lateral epicondyles of the humerus and the olecranon of the ulna. What is the relationship between these three parts in (*a*) flexion (*b*) extension of the elbow? What is the carrying angle?

2. Palpate the ulnar nerve at the elbow.

3. Mark the position of the head of the radius and the radio-humeral joint.

4. Outline the biceps muscle and show its distal tendon and the bicipital aponeurosis.

5. Show the actions of the biceps and brachialis.

6. Palpate the brachial artery and mark its course.

7. Show the position and action of the following muscles
   1. the brachioradialis,
   2. the pronator teres,
   3. the flexor muscles of the forearm,
   4. the extensors carpi radialis longus and brevis.

8. Mark the subcutaneous (posterior) border of the ulna and its styloid process and head.

9. Investigate whether the ulna moves during pronation.

10. Outline the lower end of the radius and mark its styloid process and dorsal tubercle.

11. Show the movement of the lower end of the radius during pronation.

12. By how much is one styloid process more distal than the other?

PRACTICAL CLASS 5

THE OSTEOLOGY OF THE WRIST AND HAND

Requirements: Articulated bones of the forearm and hand; skin pencils and subjects.

1. On the articulated skeleton of the hand show the scaphoid, the lunate, the triquetrum, the pisiform, the hamate, the capitate, the trapezoid, the trapezium, the metacarpal bones and the phalanges.

2. Note the attachments of the flexor retinaculum.

3. Identify the distal attachments of
   1. the flexor carpi ulnaris,
   2. the flexor carpi radialis,
   3. the extensor carpi ulnaris,
   4. the extensors carpi radialis longus and brevis.

4. Show on the wrist and hand the areas of attachment of
   1. the thenar muscles,
   2. the hypothenar muscles,
   3. the flexor pollicis longus,
   4. the flexor digitorum superficialis,
   5. the flexor digitorum profundus,
   6. the adductor pollicis,
   7. the lumbricals,
   8. the dorsal and palmar interossei.

THE ANATOMY OF THE WRIST AND HAND IN THE LIVING SUBJECT

1. Palpate the tubercle of the scaphoid, the pisiform, the tubercle of the trapezium and the hook of the hamate. Indicate the position of the lunate.

2. Demonstrate the actions of the following muscles
   1. the flexor carpi radialis,
   2. the palmaris longus,
   3. the flexor carpi ulnaris,
   4. the extensor carpi ulnaris,
   5. the extensors carpi radialis longus and brevis,
   6. the flexor digitorum superficialis,
   7. the flexor digitorum profundus,
   8. the extensor digitorum,
   9. the flexor pollicis longus,
   10. the extensor pollicis longus,
   11. the abductor pollicis longus,
   12. the adductor pollicis,
   13. the opponens pollicis,
   14. the flexor pollicis brevis,
   15. the flexor digiti minimi,

16. the opponens digiti minimi,
17. the lumbricals,
18. the palmar interossei,
19. the dorsal interossei.

3. Mark the tendons of the flexor carpi radialis, palmaris longus, and flexor carpi ulnaris. What tendons of flexor digitorum superficialis can be marked? Indicate the position of the median and ulnar nerves at the wrist.
4. Feel the radial pulse in front of the lower end of the radius.
5. Define the " anatomical snuff box " by contracting the abductor pollicis longus, the extensor pollicis brevis and the extensor pollicis longus. Can the radial pulse be felt on the floor of the snuff box?
6. Mark the distal limit of the superficial and deep palmar arches. The former is at the level of the web of the extended thumb, and the latter about 2 cm proximal.
7. Outline the synovial sheaths in the hand.
8. Note the movements of the tubercle of the scaphoid and pisiform which occur in movements of the hand at the wrist.
9. Determine the extent of the movements that are possible at the carpometacarpal joints. Which metacarpal is the most and which is the least mobile?

PART II - LOWER LIMB

CHAPTER 13

THE FUNCTIONS OF THE LOWER LIMB IN MAN

THE lower limb is concerned with the maintenance of posture and with locomotion, and its structure is an expression of these two functions. The description of the anatomy of the limb will therefore be arranged round the joints and the structure of the whole limb will be broken up into its functional components, just as the movement of the whole limb can be analysed by considering the movements of its individual segments. The dissection falls easily into regional subdivisions—the pelvic girdle, the hip joint, the knee joint, the ankle joint and the foot. In each one of these regions the student must examine the arrangement of the soft tissues around the pivots of the movements, *i.e.* the joints, in order that he may understand how such movements can be performed. To get a clear idea of the arrangement of the soft tissues, it is necessary to know about the bones of the region because muscles are attached to and act on the bones.

The description of the anatomy of each region is arranged so that the greatest attention is given to those structures that subserve the main functions of the limb, namely the muscles and the joints. Thus when dissection of the limb is finished the student should have a clear idea of the movements of each segment of the limb, but he must still integrate these separate segments into one whole, since the limb is mostly used as a single functional entity. The concluding section (Chapter 22) will show how the separate functions of the segments can be integrated, or in other words, how the muscles and the joints function in the maintenance of posture and in locomotion. Also in this concluding section will be considered the anatomy of certain structures that cannot be split up into regions but must be integrated into the concept of the entire limb—the superficial venous drainage, the cutaneous nerve supply, and the lymph drainage.

The pelvic girdle, thigh, leg and foot should be considered as a system of jointed levers serving to support and propel the weight of the body. On account of the mobility at the joints the bones seldom support the weight in stable equilibrium, *e.g.* as the legs support the top of a table to which they are firmly fixed. In the lower limb, in addition to the stresses due to the body weight, which are taken by the bones, there will continually be stresses set up by forces tending to upset the balance. These are compensated for, and stability achieved, by muscles and ligaments that will be found arranged as braces guarding each joint on all sides. A similar type of bracing is seen in some radio transmitting masts, where the long steel mast is supported on a ball and socket at its base, and has braces of long and short steel hawsers on all sides. In the case of the human lower limbs, the braces serve not only for stabilisation but, since they are mostly muscular, they move the body by their contraction. Stabilisation where there is little movement may be effected by ligaments alone and where there is free movement it is effected mainly by muscles.

Among bipedal animals the upright posture in man is unique in that the sole of the foot is flat on the ground, the foot is at right angles to the leg, and the leg, thigh and trunk are more or less vertically above each other (Chapter 22). In the anatomical position, the line of weight (the perpendicular through the centre of gravity) passes slightly behind the transverse axis of the hip joints, slightly in front of the transverse axis of the knee joints and in front of the transverse axis of the ankle joints. Strong supporting ligaments are found in front of the hip joint and behind the knee joint. Similar ligaments are not found behind the ankle joint. The upright posture in man can be maintained with a minimum of muscle effort. Variations in the upright posture are infinite in number and these can be held for a certain amount of time by contraction of the appropriate muscles. In other bipedal animals and the primates the foot is bent upwards at the ankle, and the leg and trunk are bent at the knee and hip.

Orientation and naming of parts

The **anatomical position** is one in which the person stands upright, with the feet together, the eyes looking forward, and the arms straight along the sides of the body and the palms of the hands directed forwards. The front of the body is called the **anterior** (ventral) surface and the back is called the **posterior** (dorsal) surface (front cover drawing). Higher structures are **superior** and lower structures are **inferior. Median** structures are found in the midline of the body (or of a limb) and the terms **medial** (nearer to) and **lateral** (further from) are relative to the midline.

A **sagittal plane** passes vertically anteroposteriorly through the body and movements in this plane are called **flexion** (forwards) or **extension** (backwards) (see inside front cover). However, the lower limb is twisted so that the dorsum of the foot faces forwards. This twist includes the knee. Flexion at the knee is backward movement of the leg and extension is forward movement. Downward movement of the foot and toes is **plantar flexion** (flexion) and upward movement of the same parts is **dorsiflexion** (extension). A vertical plane at right angles to the sagittal is called a **coronal** or **frontal plane.** Movement of the limb in this plane away from the midline is **abduction** and towards the midline **adduction.** At certain joints, mainly the hip joints, **rotation** also occurs about a longitudinal axis. The terms used to describe movements at the hip joint, flexion, extension, etc. have similar meanings to the same terms used elsewhere in the body.

Owing to the twisting of the lower limb, it is not so easy to identify its pre-axial and postaxial borders. The **pre-axial border** is on the lateral side of the upper part of the thigh and the medial side of the knee, leg and foot. The **postaxial border** is on the medial side of the thigh and the lateral side of the knee, leg and foot. The **ventral** and **dorsal axial lines** are on either side of a strip of skin running from the buttock down the back of the thigh and knee to the lower part of the calf. The dorsal axial line also extends horizontally for a short distance on the lateral side of the knee (indicated by heavy lines in Fig. 62).

The lower limb is attached to the vertebral column by the pelvic girdle. This consists of the two hip bones. The pelvis itself includes the sacrum which articulates with the hip bones at the sacro-iliac joints. These are almost immobile and very stable. The limb consists of three segments, the thigh from the hip to the knee, the leg from the knee to the ankle, and the foot. The foot has plantar and dorsal surfaces and five toes. The big toe (the hallux) is medially placed. There are two segments in the big toe and three in the other toes, though occasionally the distal two phalanges in the little toe are fused together.

CHAPTER 14

THE PELVIC GIRDLE

INTRODUCTION

WEIGHT is transferred from the vertebral column to the sacrum at the lumbosacral joints and from the sacrum to the two hip bones through the sacro-iliac joints. Since very little movement occurs at the latter joints they are stabilised mainly by ligaments. Weight is transferred partly across the interlocking joint surfaces and partly across a system of ligaments by which the sacrum is, as it were, slung between the two hip bones.

The main line of weight-bearing in the hip bone runs from the sacro-iliac joint to the acetabulum and the bone is thickest along this line. The whole arrangement may be considered as an arch of which the sacrum is the keystone, and these thickened regions of bone are the pillars. The weight is continued downwards to the femora. The symphysis pubis is thus a tie, preventing the arch from spreading apart. The analogy is very imperfect, however, since the sacrum is partly slung from the ilia and thus may act as a movable keystone.

The hip joint is a ball and socket joint allowing a wide range of movement. The body is balanced on it, with the help of a large number of ligaments and muscles acting as braces. Some of the muscles are short whereas the longer ones exert considerable leverage and produce powerful and extensive movements.

On the skeleton, note, on the back of the sacrum, the **median sacral crest** in the midline and the **dorsal** (posterior) **sacral foramina.** On the front of the sacrum identify the **pelvic** (anterior) **sacral foramina** and the bodies and lateral parts of the individual vertebral components of the sacrum. Examine the iliac surface of the bone and note the ear-shaped (auricular) articular surface, and the rough non-articular area behind for the attachment of the interosseous and dorsal sacro-iliac ligaments. Passing from the back of the sacrum to the tuberosity and the spine of the ischium are the **sacrotuberous** and **sacrospinous ligaments** respectively.

On the hip bone, note the surface for articulation with the sacrum and the large area for the attachment of the interosseous and dorsal sacro-iliac ligaments. The thin ventral sacro-iliac ligament is attached in front of the joint, and above, the iliolumbar ligament passes from the transverse process of the 5th lumbar vertebra to the neighbouring part of the ilium. The **symphysis pubis** is the joint between the two pubic bones and has a capsule with supporting ligaments and a fibrocartilaginous disc between the two bony surfaces.

DISSECTION

If the pelvis has been dissected, remove the structures in front of the sacro-iliac joint, cut through the ventral ligament and separate the joint surfaces. Do not cut the interosseous and dorsal ligaments at this stage but note their thickness. Examine the structure of the symphysis pubis. Note that the surfaces of the pubic bone are covered by hyaline cartilage and that there is a mass of fibrocartilage between them. If the pelvis has not been dissected, leave the inspection of the sacro-iliac joints and the symphysis pubis until later (Vol. I, page 147).

STRUCTURAL DETAILS

The **sacro-iliac joint,** although a synovial joint, does not permit much movement because of the interlocking depressions and elevations on the corresponding cartilaginous surfaces of the bones. The cartilage covering the surfaces may disappear in places with advancing years and be replaced by a fibrous union which limits movement. Immediately behind the cartilaginous surfaces, the bones are united by thick **interosseous ligaments.** The dorsal and interosseous sacro-iliac ligaments are very strong and form the sling which carries the sacrum and its superimposed weight on the hip bones. There is a tendency for the top of the sacrum to tilt downwards and forwards into the pelvis. The corresponding backward and upward movement of the lower part of the sacrum and coccyx is prevented by the sacrospinous and sacrotuberous ligaments.

The **symphysis pubis** is between the bodies of the two pubic bones. The joint surfaces are covered with hyaline cartilage and between them is a disc of fibrocartilage. The capsule of the joint is not strong.

Towards the end of pregnancy some increased movements at the joints of the pelvic girdle may occur and permit a slight increase in the circumference of the pelvis. This may be important during childbirth. Bone may be absorbed, especially in the region of the symphysis pubis, but later it is replaced.

CHAPTER 15

THE LATERAL AND POSTERIOR ASPECTS OF THE HIP JOINT

INTRODUCTION

IN this region, the largest structures are the gluteal and hamstring muscles, which are strong extensors of the thigh at the hip joint. The hamstrings, however, are also flexors of the leg at the knee. The sciatic nerve, the largest nerve in the body, is a prominent structure in this region.

The pelvic bones were examined during the dissection of the abdomen and pelvis. The principal bony landmarks were identified (Vol. I, page 78), but special attention should now be paid to the outer surface of the hip bone. The outer surface of the ala of the ilium is divided into a small posterior area, a large anterosuperior area and a large antero-inferior area, and these areas serve for the attachment of gluteus maximus, medius and minimus respectively. Identify the outer surface of the **ischiopubic ramus,** to which the adductor muscles are attached, and the **ischial tuberosity,** where the hamstring muscles arise. Above the obturator foramen, find the **acetabulum** for articulation with the head of the femur. The articular surface is horseshoe-shaped and the acetabular edge is incomplete below (the **acetabular notch**).

The upper end of the femur consists of a **head,** which articulates with the acetabulum, a **neck** and a **body** on which are the two trochanters. The **lesser trochanter** is medial and inferior and the **greater** is lateral and superior. Joining the trochanters are the **intertrochanteric crest** behind and the **intertrochanteric line** in front. The body of the bone is smooth except for a vertical ridge, the linea aspera, on its posterior surface.

DISSECTION

Make incisions along the whole length of the lateral and medial sides of the thigh as far as the knee (back cover). Reflect the two flaps of skin downwards. Remove the subcutaneous tissue from the buttock and the back of the thigh. Note the toughness and thickness of the dep fascia of the thigh (the **fascia lata**) and of the gluteal region. It is particularly thick on the outer side and is here known as the **iliotibial tract,** which can be followed down as far as the lateral condyle of the tibia. Note the upper attachments of the large **gluteus maximus muscle** to the posterior area on the outer surface of the ala of the ilium, to the back of the sacrum and to the sacrotuberous ligament. It is attached below partly to the gluteal tuberosity on the body of the femur between the greater trochanter and the linea aspera, but mostly to the iliotibial tract. Further forwards, note another smaller muscle also attached to the iliotibial tract. This is the **tensor fasciae latae** which arises from the iliac crest in front.

Clean the surface of the gluteus maximus and then cut through the muscle at right angles to its fibres, one third distal to its upper attachment. Reflect its two parts medially and laterally, cutting any branches of the inferior gluteal vessels and nerve which supply it. The deeper part of the gluteal region is now exposed (Fig. 26).

Begin by identifying the **piriformis muscle.** It arises inside the pelvis on the front of the sacrum and passes into the thigh through the greater sciatic foramen to be attached to the upper edge of the greater trochanter. Clean its surface and its upper and lower edges. Above the upper edge of the piriformis, branches of the superior gluteal artery and nerve are found emerging from the pelvis. Trace the branches of the nerve forwards and upwards between the gluteus medius and minimus muscles. These muscles arise from the outer surface of the ilium, and pass downwards to their attachments to the greater trochanter. Clean them and dissect their tendons to their insertions. Cut across the medius to expose the minimus.

The **gluteus medius** is attached above to the upper outer surface of the ilium and below to the lateral surface of the greater trochanter. It is the most important lateral brace and abductor of the hip joint, preventing the body from falling to the unsupported side when the weight is carried on one lower limb. The **gluteus minimus** is attached above to the lower outer surface of the ilium and below to the anterior surface of the greater trochanter. It is an abductor and medial rotator of the thigh. The superior gluteal nerve supplies these two muscles and the tensor fasciae latae.

At the lower border of the piriformis find the large **sciatic nerve,** the largest motor and sensory nerve of the lower limb. Clean it for

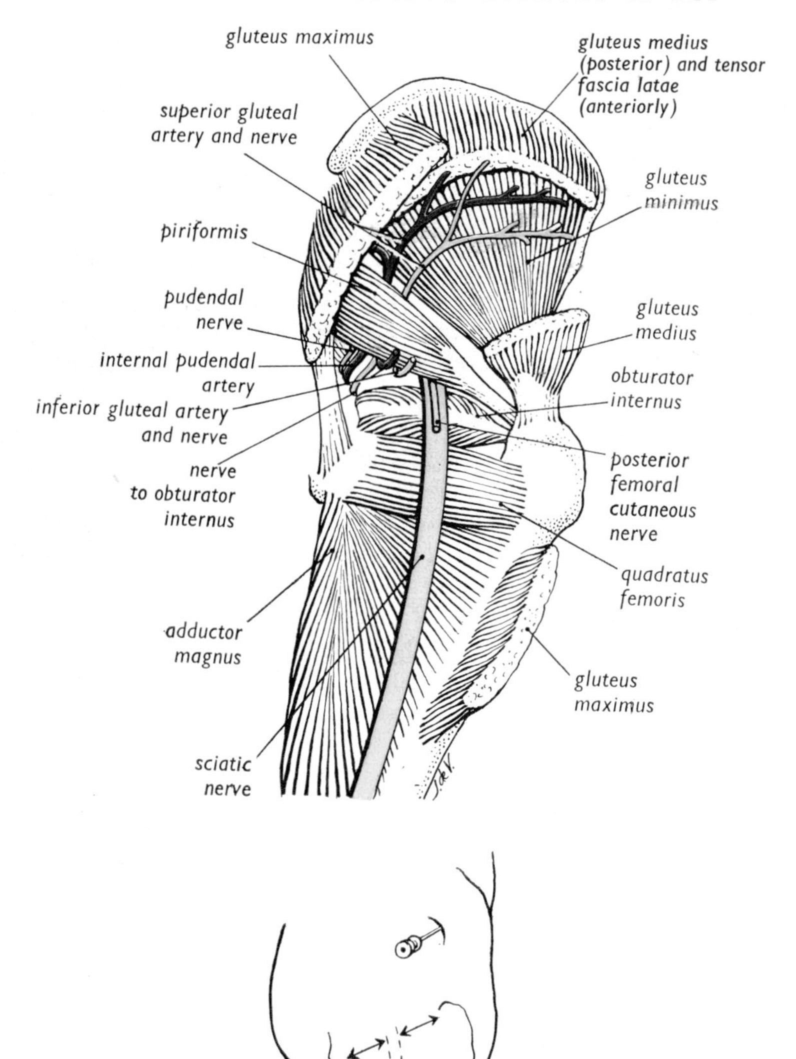

FIG. 26
Diagram of the gluteal region. The lower figure indicates the position of the sciatic nerve and also a site commonly used for intramuscular injections.

a short distance down the thigh. A large sensory nerve, the **posterior femoral cutaneous nerve,** is found lying superficial to it. On the medial side of the sciatic nerve are branches of the inferior gluteal artery and vein and the stump of the inferior gluteal nerve. This nerve was cut during the reflection of the gluteus maximus, which it supplies. Deep to the sciatic nerve and just inferior to the piriformis find the tendon of the **obturator internus muscle.** Trace the tendon of the obturator internus medially. It can be followed through the lesser sciatic foramen into the pelvis where it is attached to the inner surface of the obturator membrane and the surrounding bone. Laterally the tendon is attached to the upper edge of the greater trochanter. The piriformis and the obturator internus are both lateral rotators of the thigh.

Identify the edge of the large **sacrotuberous ligament.** Remove the fibres of the gluteus maximus and cut the ligament away in order to expose more fully the back of the ischial spine. Note the **sacrospinous ligament** attached to the spine. Dissect carefully over the back of the spine and adjacent portion of the ligament. Identify, from lateral to medial, the nerve to obturator internus, the internal pudendal artery and vein and the pudendal nerve, passing over the back of the spine and ligament. Note the position of the quadratus femoris muscle below the obturator internus, clean its posterior surface and identify, at its lower border, the upper border of the adductor magnus muscle. The **quadratus femoris** runs between the ischial tuberosity and the intertrochanteric crest. It is a lateral rotator of the thigh.

On the back of the thigh are long muscles that act at the hip and the knee joints. Identify the three **posterior femoral (hamstring) muscles,** arising from the ischial tuberosity (Fig. 27). Follow them downwards and note that the **biceps** lies laterally and the **semitendinosus** and **semimembranosus** medially. The semitendinosus lies on the posterior surface of the semimembranosus. Clean all three down to the level of the knee joint. They all assist the gluteus maximus and act as posterior braces of the hip, preventing the body from falling forwards. When the lower limb is free they pull it backwards. The three hamstring muscles are also powerful flexors at the knee joint. Examine the posterior surface of the **adductor magnus;** clean it and the sciatic nerve which lies on it. The posterior

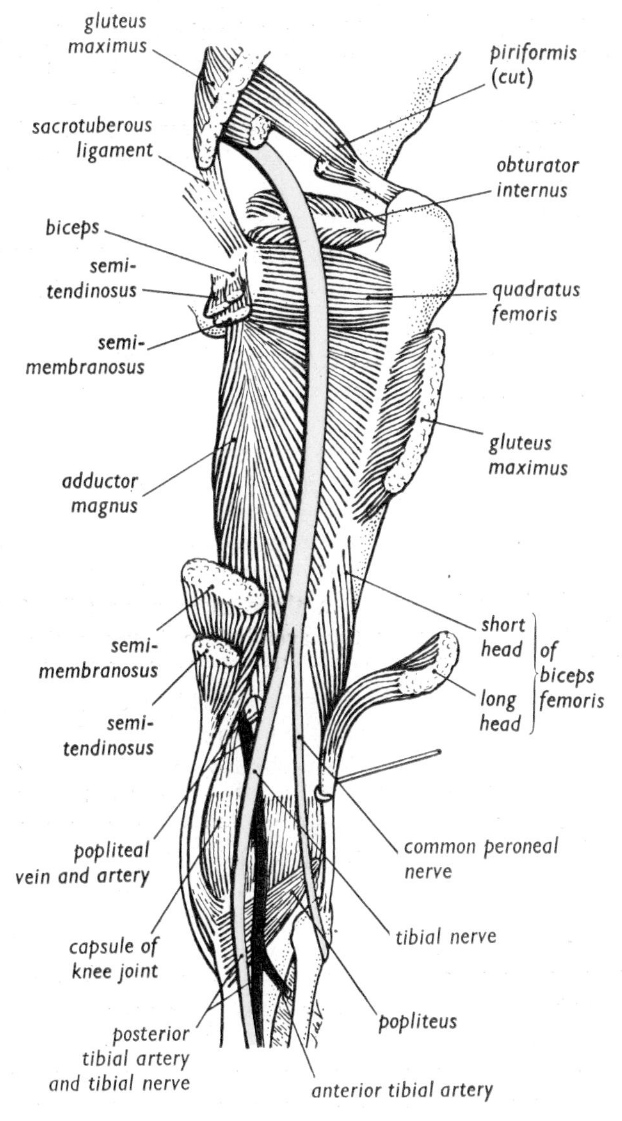

FIG. 27
The course and important relations of the sciatic nerve

fibres of the adductor magnus from the ischial tuberosity are attached below to the adductor tubercle of the femur. The biceps femoris has a second head of attachment from the linea aspera, and the whole muscle is attached to the upper end of the fibula. The semitendinosus and semimembranosus pass along the inner side of the knee to be attached to the upper medial part of the tibia. Find the branches of the sciatic nerve to the hamstring muscles and note its division into the tibial and common peroneal nerves. Occasionally this division occurs in the pelvis in which case the common peroneal nerve passes through the piriformis.

STRUCTURAL DETAILS

The hip bone (Vol. I, page 78)

The inner and outer aspects of the hip bone have already been described. The three bones forming the hip bone are modified for weight-bearing or for muscular attachment (ischial tuberosity). Parts of the bone where there are few stresses are thin (ala of the ilium) or replaced by membrane (obturator foramen).

The hip bone is **ossified** in three main parts (Fig. 28). All three centres appear near the acetabulum between the second and fifth months of intra-uterine life and in the following order, ilium, ischium, pubis. The ilium and ischium each form about two-fifths of the acetabulum and the pubis one-fifth. The Y-shaped cartilage between the three bones begins to ossify at about the twelfth year and ossification is complete by the fifteenth year. There are secondary centres for the iliac crest, anterior inferior iliac spine, ischial tuberosity and symphysis pubis. They appear at about puberty and fuse at about the twenty-fifth year.

The femur

The body of the femur can be thought of as an arched tubular pillar which carries the weight of the body. This is mechanically effective because no stresses fall on the hollow centre of such a pillar. The greatest stresses during weight-bearing occur in the head, neck and upper third of the body. Though the bone is thickened and the trabeculae of bone are arranged to meet this stressing, the neck of the femur is frequently fractured.

The various lines and elevations on the bone are all connected with the pull of attached muscles or ligaments, and many of them only become fully developed under the influence of this pull. The body has a ridge down the back called the linea aspera (Fig. 29). Elsewhere the body is smooth. The upper end of the femur has a rounded head for articulation with the hip bone and two prominences for muscular attachments—the greater and lesser trochanters. The two trochanters are joined in front by the intertrochanteric line and behind by the intertrochanteric crest. On the medial

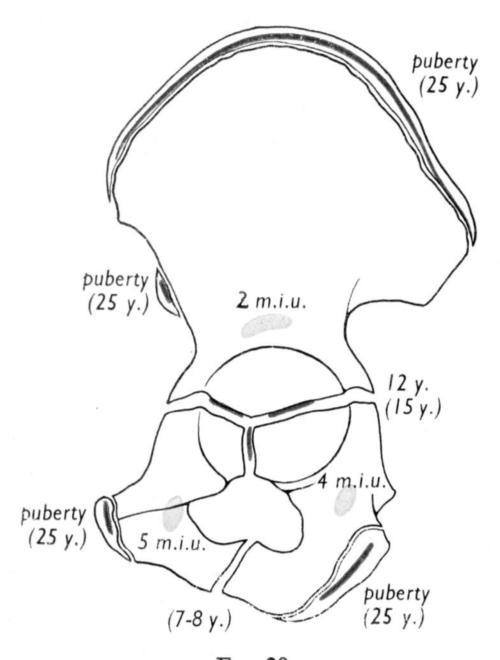

FIG. 28

The primary centres of ossification of the hip bone are shown in yellow and the secondary centres in red. The figures in brackets indicate the times of fusion of the primary and secondary centres (y., years; m.i.u., months in utero).

aspect of the head is a depression for the attachment of the ligament of the head of the femur. Between the head and the body is the neck, which is directed medially, upwards and forwards.

The femur has a primary centre of **ossification** for the body which appears at about the end of the second month of intra-uterine life, and secondary centres for the head (first year), for the greater trochanter (fifth year) and the lesser trochanter (tenth year). The

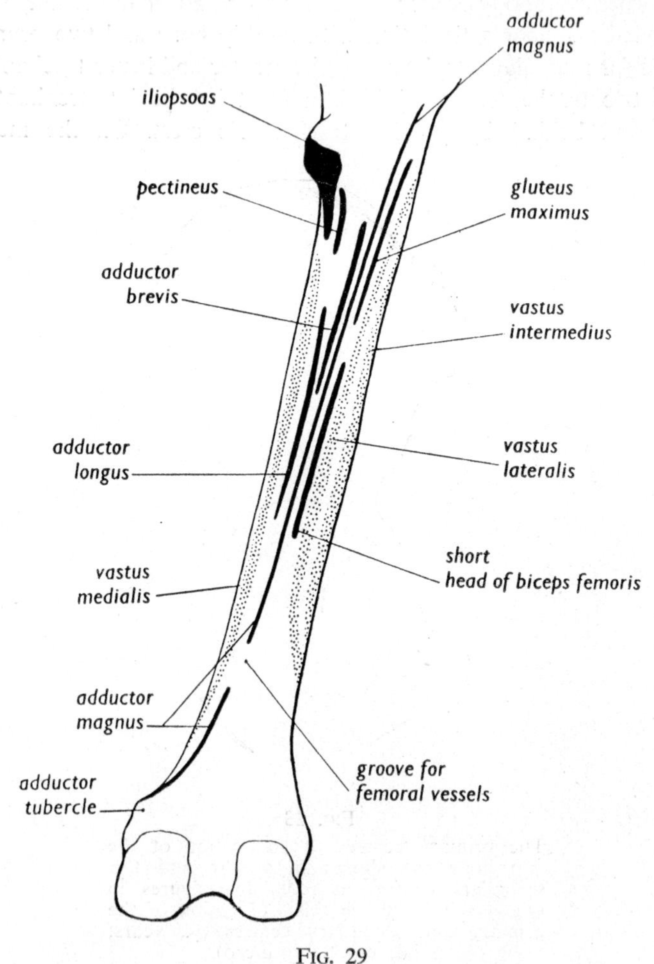

FIG. 29
Diagram of the muscular attachments on the back of the femur. The width of the linea aspera has been much exaggerated.

upper epiphyses join the body at about the eighteenth year, the last to do so being the head. The neck is ossified as part of the body (Fig. 30).

The gluteus maximus

This muscle is a posterior brace of the hip joint, preventing the body from falling forwards if the weight passes in front of the head of the femur. It is therefore the muscle specially concerned in attaining the characteristic upright posture, and it plays a large part in raising the body from the stooping position and when climbing stairs or a hill. When the limb is swinging freely the gluteus maximus pulls it backwards and it is therefore an extensor of the thigh. Through its large attachment to the iliotibial tract, the muscle has an important action on the knee joint where it acts as an extensor of the tibia on the femur if the foot is fixed on the ground.

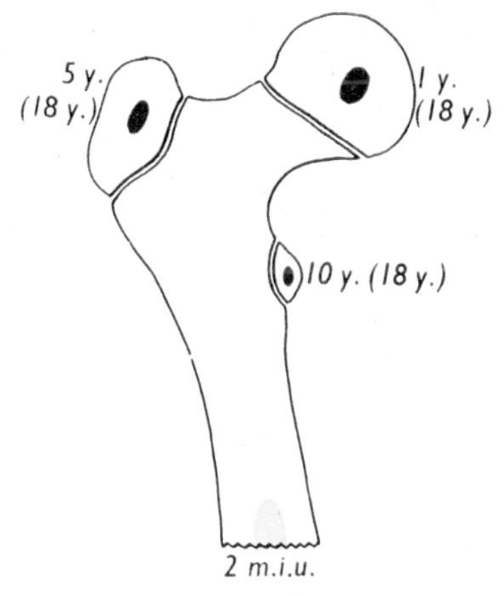

FIG. 30
The primary centre of ossification for the upper end of the femur is shown in yellow and the secondary centres in red. The figures in brackets indicate the times of fusion of the primary and secondary centres (y., years; m.i.u, months in utero).

The tensor fasciae latae

Tensor fasciae latae (an abductor) acts as a lateral brace, helping to prevent the body from falling to the opposite side when the weight is balanced on the head of the femur. Several other muscles help in this important action. It is also a flexor and medial rotator of the thigh. The tensor fasciae latae also produces extension at the knee joint if the foot is fixed on the ground, through the iliotibial tract which is a specially thickened portion of the fascia lata. The whole of the tensor fasciae latae and most of the gluteus maximus are attached to the tract. The fascia lata is the deep fascia of the leg, and its lateral thickening, the iliotibial tract, has a strong attachment to the outer aspect of the upper end of the tibia. Above, the tract is attached to the iliac crest.

The gluteus medius and minimus

Gluteus medius is an important muscle of the hip joint. It is an abductor of the pelvis on the femur. In walking, as the weight passes on to the foot that is on the ground, the gluteus medius of that side contracts, prevents the unsupported side of the pelvis from falling and slightly raises the pelvis on the femur (Fig. 35). This helps the non-weightbearing limb to swing forward. If the right gluteus medius is paralysed, standing on the right lower limb results in the left side of the pelvis falling. Gluteus minimus assists the medius in abduction of the thigh at the hip joint and is also an important medial rotator at this joint.

The **piriformis, obturator internus** and **quadratus femoris muscles** are lateral rotators of the femur at the hip joint. These short muscles and their tendons are in intimate contact with the joint capsule and act as stabilisers of the back of the joint.

The sciatic nerve

The sciatic nerve is formed by branches from the anterior primary rami of the 4th and 5th lumbar and 1st, 2nd and 3rd sacral spinal nerves, and enters the thigh at the lower border of the piriformis (Fig. 27). It curves slightly laterally before running down the middle of the back of the thigh. Usually about half-way down the back of the thigh it divides into the common peroneal (lateral) and tibial (medial) nerves. Its posterior relations are from above downwards, the gluteus maximus and the biceps. It is also overlapped to a slight extent by the semimembranosus. Anteriorly are the body of the ischium, the tendon of the obturator internus which separates it from the hip joint, and below this the quadratus femoris and the adductor magnus. As the sciatic nerve enters the thigh, the posterior femoral cutaneous nerve is posterior, and on its medial side are first the inferior gluteal vessels and nerve, and then more medially are the nerve to obturator internus, the internal pudendal vessels and the pudendal nerve. At the lower edge of the gluteus maximus and lower down between the biceps and semitendinosus the sciatic nerve is covered only by fascia, fat and skin.

The **cutaneous nerves of the gluteal region** are derived from the dorsal rami of lumbar and sacral spinal nerves (Chap. 23), together with branches from the iliohypogastric and the lateral and posterior femoral cutaneous nerves.

CHAPTER 16

THE ANTERIOR AND MEDIAL ASPECTS OF THE HIP JOINT

INTRODUCTION

THE front of the thigh contains two groups of muscles which are roughly separated by the line of the sartorius muscle passing obliquely downwards and medially to the inner aspect of the knee. Medial to the sartorius, the muscles act predominantly on the hip joint, and those lateral to the sartorius act on the knee joint. The latter group of muscles, quadriceps femoris, will be considered with the knee joint but one part of it, rectus femoris, acts on both the hip and the knee joints. The muscles to be considered now are mostly flexors, medial rotators and adductors of the femur at the hip joint.

On the articulated skeleton, identify the lumbar vertebral bodies and transverse processes, the ala of the ilium, the ischiopubic ramus, the pubic tubercle, the lesser trochanter and the linea aspera.

DISSECTION

The **femoral triangle** is formed by the inguinal ligament above, the sartorius laterally, and the adductor longus medially. In the superficial fascia towards the medial side of the thigh is found the large, subcutaneous **great saphenous vein,** running upwards from the lower part of the limb and piercing the deep fascia to reach the femoral vein. The vein pierces the deep (cribriform) fascia at the **saphenous opening,** which is 4 cm below and lateral to the pubic tubercle. Define the margins of the saphenous opening by removing the loose areolar tissue covering it. A few enlarged lymph nodes may be found lying alongside the vessels and also just distal to the inguinal ligament. Define the **sartorius muscle** running from the anterior superior iliac spine across both hip and knee joints to the medial side of the tibia. It is supplied by the femoral nerve. While removing the sheet of fascia covering the triangle, cutaneous branches of the femoral nerve will be encountered. Their distribution need not be studied at present, but they should be traced upwards to their origin in order to find and identify their parent trunk, the **femoral nerve.** Passing down the middle of the femoral triangle are the femoral artery and vein.

Examine the **femoral sheath** surrounding the artery and vein. Deep to the inguinal ligament it is continuous with the extraperitoneal fascia on the anterior and posterior abdominal walls (Fig. 31).

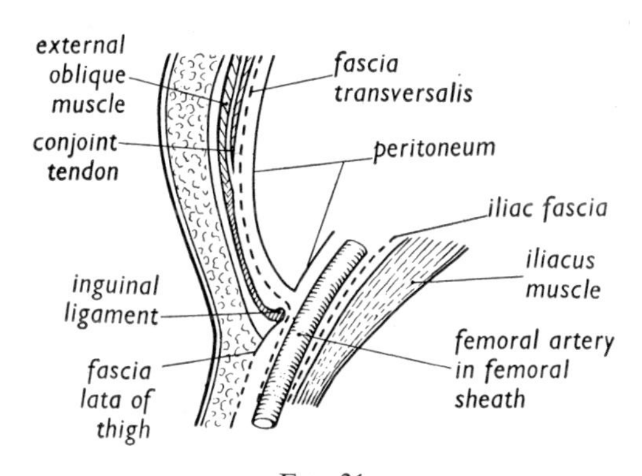

FIG. 31
Diagram of the formation of the femoral sheath.

There are three compartments in the sheath, one for the artery laterally, one for the vein in the middle, and one medially, the **femoral canal** (Fig. 32). Open the sheath by an anterior incision and insert the little finger into the canal. Push the finger upwards deep to the inguinal ligament and note that the canal opens above into the abdominal cavity, external to the peritoneum. Examine the relations of this opening (the **femoral ring**); laterally is the femoral vein, medially the sharp lateral edge of the lacunar ligament, anteriorly the inguinal ligament, and posteriorly the pectineal ligament. These relations are of importance as the femoral canal may be the site of a femoral hernia. The femoral nerve lies outside the sheath lateral to the artery.

Clear away the fascia round the femoral vessels and nerve and examine the muscles forming the floor of the triangle and the relations of the femoral vessels. The **femoral artery** gives off the following branches—the medial and lateral femoral circumflex arteries and the profunda femoris artery, running respectively backwards, laterally and downwards to supply the deeper parts of the thigh. Examine the point at which the femoral artery passes deep to the inguinal ligament and determine that it lies midway between the

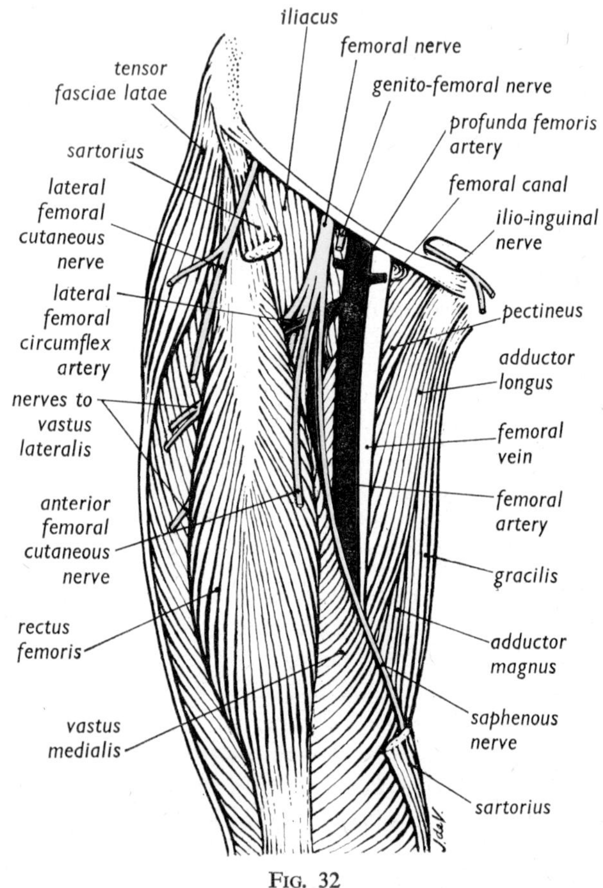

FIG. 32
Diagram of the principal relations of the structures on the front of the thigh.

anterior superior iliac spine and the pubic symphysis. This provides a surface marking for the femoral artery. The floor of the triangle is formed, from the lateral to the medial side, by the **ilio-psoas, pectineus** and **adductor longus muscles.** Pass the handle of the scalpel into the interval between the pectineus and the iliopsoas and note that the scalpel meets the anterior surface of the hip joint. Deep to the upper part of the sartorius and lateral to the tendon of the iliopsoas is the tendon of the **rectus femoris.** This muscle forms part of the quadriceps femoris and its upper attachments are to the anterior inferior iliac spine and to the upper edge of the acetabulum.

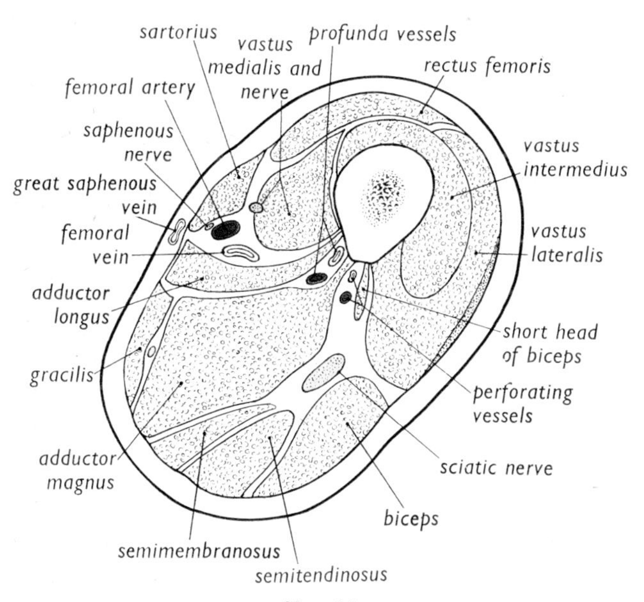

FIG. 33
Drawing of a cross section of the middle third of the thigh.

The vessels beyond the triangle lie in the **adductor canal** (Hunter's canal), an intermuscular gap formed by the sartorius superficially, the adductors posteriorly and the vastus medialis laterly (Fig. 33). Lift up the sartorius muscle and clear away the fascia deep to it to a point just above the medial side of the knee. Take care, however, to preserve the femoral artery and vein and two nerves, the saphenous nerve and the nerve to the vastus medialis (branches of the femoral nerve). Follow the vessels to a point approximately two-thirds of the way down the thigh, where they pass through the adductor magnus muscle to reach the popliteal fossa behind the knee joint. The saphenous nerve is found to leave the vessels and pass through the roof of the canal towards the medial side of the knee where it becomes superficial.

The dissection of the medial side of the thigh must now be completed. Find the **gracilis,** the most medial muscle of the thigh. It is attached above to the ischiopubic ramus and below to the upper part of the medial surface of the tibia. Examine the attachment of the **adductor longus muscle** to the front of the body of the pubis and then cut through the muscle about 1 cm away from the bone. Turn the muscle downwards, exposing the **adductor brevis muscle.** Clean the front surface of this and find the anterior branch of the **obturator nerve** and the profunda femoris artery, which cross anterior to it. The adductor brevis is attached to the ischiopubic ramus. Cut through the adductor brevis half-way along its length, pull aside the two portions of the muscle and expose the front surface of the **adductor magnus muscle.** Lying between the brevis and magnus is the posterior branch of the obturator nerve. The adductor magnus is attached to the ischiopubic ramus and to the ischial tuberosity. The adductor muscles (except the gracilis) are attached to the femur along the linea aspera (Fig. 29). The adductor magnus has the longest attachment, extending between the greater trochanter above and the adductor tubercle below.

These muscles are all adductors and lateral rotators of the thigh. With pectineus, they are the medial braces of the hip joint. They are used particularly in crossing the thighs or grasping with the thighs, as in horse riding.

STRUCTURAL DETAILS

The flexors at the hip joint

The iliopsoas, rectus femoris and pectineus act as braces to the front of the joint, that is to say they prevent the body falling backwards on the head of the femur or they flex the femur on the hip bone, as in walking and in raising the lower limb when lying on the back or climbing stairs. The upper end of the **rectus femoris** is attached to the anterior inferior iliac spine and to the upper edge of the acetabulum. The **iliacus** and **psoas,** the strongest flexors, have already been seen in the abdomen, where their upper attachments were examined. Their combined tendon is attached to the lesser trochanter. The **pectineus** runs from the superior ramus of the pubis, lateral to the pubic tubercle, downwards to the femur and is attached to the back of the bone below the lesser trochanter. The psoas is supplied by the ventral rami of the lumbar nerves, and the iliacus, pectineus and rectus femoris by the femoral nerve.

The adductors at the hip joint

The adductor group of muscles lies on the medial side of the thigh. The **adductor magnus** has the longest attachments both to the ischiopubic ramus and to the linea aspera. The fibres which come from the most anterior part of the ischiopubic ramus are inserted highest on the linea aspera. The most posterior fibres come from the ischial tuberosity and are attached to the adductor tubercle above the medial epicondyle of the femur. The fibres in this latter part of the muscle act like a hamstring muscle as an extensor at the hip joint and frequently are supplied by the sciatic nerve. The other adductor muscles, longus, brevis and gracilis, are much smaller. They, and the magnus, are supplied by the obturator nerve.

The femoral vessels

The **femoral artery,** the continuation of the external iliac artery, enters the thigh by passing deep to the inguinal ligament halfway between the symphysis pubis and the anterior superior iliac spine. It passes through a hole in the adductor magnus close to the femur, two-thirds of the way down the thigh. In the upper part of its course it runs from the base to the apex of the femoral triangle. The femoral vein lies medial to the artery in its upper part. The femoral nerve enters the thigh lateral to the artery. Lower down, the saphenous nerve lies on the lateral side of the artery. In front of the artery lies the deep fascia forming the roof of the femoral triangle. In the adductor canal the sartorius is anterior and medial to the artery, and the saphenous nerve crosses it from its lateral to its medial side. Posteriorly lies the femoral vein separating it from the adductor longus, and on the lateral side lie the vastus medialis and its nerve.

The femoral artery gives off three main branches, the medial and the lateral femoral circumflex and the profunda femoris arteries.

The mode of origin of these vessels varies. They supply adjacent muscles and the structures on the medial side and back of the thigh. The **femoral vein** accompanies the artery. It receives deep tributaries corresponding to the large arterial branches and also a large superficial tributary, the great saphenous vein, which passes through the saphenous opening. These large veins have valves.

The femoral nerve

The femoral nerve is formed from the ventral rami of the 2nd, 3rd and 4th lumbar spinal nerves, and after a course on the posterior abdominal wall, enters the thigh about 1 cm lateral to the femoral artery. It supplies the flexors of the femur at the hip joint and the extensors of the leg at the knee joint as well as the joints themselves. Shortly after entering the thigh it divides into a number of branches some of which pass to the skin on the front of the thigh. The deeper branches are motor to the four muscles of the quadriceps, and include an important cutaneous nerve, the saphenous, which passes downwards and supplies the skin of the medial side of the leg, ankle joint and foot as far as the ball of the big toe. The femoral nerve also gives branches to the sartorius, iliacus and pectineus muscles.

The obturator nerve

The fibres of the obturator nerve come from the ventral rami of the 2nd, 3rd and 4th lumbar spinal nerves and, after a course on the side wall of the pelvis, the nerve enters the thigh through the upper and anterior part of the obturator foramen. It divides into two branches, passing in front of and behind the adductor brevis muscle and supplies the adductors, the gracilis, the obturator externus, the hip joint, a cutaneous branch to the medial side of the thigh, and a small branch to the knee joint.

FUNCTIONAL ASPECTS

Femoral hernia

Occasionally a process of peritoneum and some of the abdominal contents pass down the femoral canal, enter the thigh, and form a femoral hernia. The hernia becomes subcutaneous by protruding through the saphenous opening. A femoral hernia lies below and lateral to the pubic tubercle and an inguinal hernia lies above the tubercle. A femoral hernia occurs more frequently in women than in men, possibly owing to the greater space between the inguinal ligament and the superior pubic ramus.

Referred pain

The femoral, the sciatic and the obturator nerves give branches to the hip joint as well as to the knee joint. Because of this, pain originating in the hip may be referred to the knee.

Varicose veins

The large superficial veins are connected with the deep veins by vessels which perforate the deep fascia. If the valves of the superficial and perforating veins become incompetent then back pressure causes the unsupported veins outside the deep fascia to become dilated, elongated and tortuous (varicose).

CHAPTER 17

THE HIP JOINT—STRUCTURE AND MOVEMENTS

INTRODUCTION

THIS ball and socket joint has to carry the weight of the body and also remain stable during extensive movements. It is surrounded by muscles, some of which are very powerful. Accidental dislocation of the joint is unusual but dislocation is found as a congenital abnormality.

Examine the acetabulum of the hip bone and note the notch, and the articular and nonarticular surfaces. On the head of the femur note the extent of the articular surface and the presence of the small non-articular fossa for the attachment of the ligament of the head of the femur. The head is joined to the body by the neck which has many foramina for vessels directed medially towards the head.

DISSECTION

Cut through the gluteus minimus, 5 cm from the femur, thereby exposing the upper aspect of the hip joint. Posteriorly cut through the piriformis and the obturator internus close to their attachments to the femur and the quadratus femoris through its middle. Pull the muscles aside and expose the back of the joint with the tendon of obturator externus running across it from below. It is attached to the fossa on the medial aspect of the greater trochanter and to the outer surface of the obturator membrane. The muscle is a lateral rotator of the thigh. Cut through the obturator externus, the adductor magnus close to its origin from the ischiopubic ramus, and finally, through the iliopsoas and pectineus about 5 cm distal to the inguinal ligament. Pull aside all the stumps of the muscles and tendons and examine the hip joint.

Look at the front of the joint. An opening is sometimes seen in the capsule, somewhat to the medial side. This, when present, is the communication between the cavity of the joint and a bursa deep to the iliopsoas tendon. With the bones in front of you, examine the attachments of the capsule to the margin of the acetabulum and to the upper end of the femur. Define a marked thickening anteriorly, running from the anterior inferior iliac spine and bifurcating before its attachment to the upper and lower ends of the intertrochanteric line. This is the **iliofemoral ligament.** Further medially, identify a thinner set of fibres running from the pubic part of the acetabular edge downwards and laterally to the lower half of the capsule in front of the joint. This is the **pubofemoral ligament.** On the back identify a third capsular thickening, running from the ischial part of the acetabular edge spirally upwards and laterally, the **ischiofemoral ligament.** Note that the direction of all these ligaments is such that they become taut in extension of the thigh.

Open the joint by cutting round the capsule on all sides. Other structures still passing between the pelvis and the lower limb should be cut as high up as possible. Disarticulate the femur from the acetabulum. While doing this, the ligament of the head of the femur, passing between the head of the femur and the acetabular notch, has to be cut. Now examine the cut surface of the capsule and feel the thickness of the tough, large, iliofemoral ligament. Note that at the attachment of the capsule to the femur certain fibres turn back along the neck. These are the **retinacula** and convey blood vessels to the neck and head of the femur. Attached to the edges of the acetabulum and the transverse ligament, which bridges the acetabular notch, is the **acetabular labrum.** This forms a complete ring round the head of the femur lateral to its equator.

STRUCTURAL DETAILS

The hip joint is a synovial joint of the ball and socket type. The ball is the rounded head of the femur, the socket is formed by the acetabulum and labrum. In comparison with the shoulder joint, the hip joint is much more stable but less mobile.

The head of the femur forms two-thirds of a sphere and is slightly flattened superiorly. It is directed upwards, forwards and medially so that in the anatomical position the anterior part of the articular surface is not in contact with the acetabulum. The articular surface of the acetabulum is horse-shoe shaped. The non-articular area is inferior and covered by a fatty pad. The rim of the acetabulum is completed across the notch by the **transverse acetabular ligament.** Both joint surfaces are covered by articular cartilage. At the edge of the acetabulum there is a ring of fibro-cartilage, the **acetabular labrum** which deepens the socket and grasps the head of the femur.

The **capsule** of the joint is very strong. It surrounds the joint like a sleeve and is attached medially to the margin of the acetabulum, to the transverse ligament and to the labrum. Laterally it is attached in front to the intertrochanteric line and at the back to the neck of the femur about 1 cm medial to the intertrochanteric crest, so that part of the posterior surface of the neck is extracapsular. Some of the capsular fibres turn back along the neck of the femur as the retinacula. The upper end of the femur has three epiphyses, one for the head and one for each of the trochanters; the former is intracapsular and the latter two extracapsular.

There are three main **capsular ligaments.** The **iliofemoral** is the largest and the most important. It is a broad powerful band attached to the pelvis just below the anterior inferior iliac spine. It divides into two limbs attached to the upper and lower parts of the intertrochanteric line. The **pubofemoral ligament** is attached to the pubic part of the acetabular margin and passes downwards to blend with the capsule in front and below. The **ischiofemoral ligament** arises from the ischial part of the acetabular margin; its fibres pass across the back of the femoral neck to the upper part of the capsule. It should be noted that these three ligaments and the fibres of the capsule are arranged spirally round the joint as they pass laterally, so that extension is checked and they are responsible for the relatively small range of that movement.

Inside the capsule is the **ligament of the head of the femur,** a flattened band, attached to the transverse ligament and to the pit on the femoral head. It plays little part in stabilising the joint but conveys blood vessels to the head of the femur when its epiphyseal cartilage is still present.

The **synovial membrane** lines the capsule and continues back on to the neck as far as the articular cartilage. The bursa under the iliopsoas tendon usually communicates with the synovial cavity of the joint.

The **nerve supply** of the joint comes, as in all joints, from the nerves supplying the muscles acting on the joint (Hilton's law). These are the femoral, sciatic and obturator nerves, and the fibres are derived from the 2nd, 3rd and 4th lumbar nerves. Therefore it is possible for pain produced by disease involving these rami

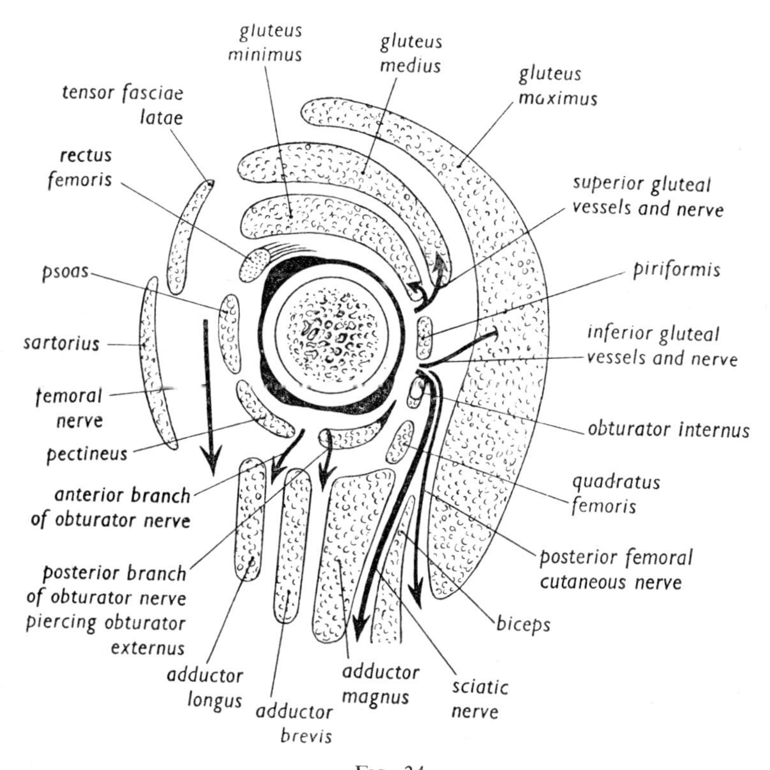

FIG. 34
Diagram of the principal relations of the hip joint. The capsular thickenings are shown but not labelled.

elsewhere (*e.g.* in the spine) to be referred to the hip joint. The **blood supply** to the joint comes from neighbouring vessels. The retinacula carry vessels to the intracapsular parts of the femur, especially the head. In some fractures of the neck of the femur, the blood supply to the head may be severed.

The **relations** of the joint are as follows (Fig. 34). Anteriorly, in front of the head of the femur, is the iliopsoas tendon and medial to this, the pectineus. In front of the iliacus is the femoral nerve, in front of the psoas are the femoral artery and vein, and in front of the pectineus is the femoral canal. Laterally and above is the rectus femoris, with the gluteus medius and minimus a little further posterior. Inferiorly is the obturator externus muscle, the tendon of which runs below and behind the neck. Posteriorly are the piriformis and below it the obturator internus and the upper edge of the quadratus femoris. Behind the latter two muscles is the sciatic nerve overlapped by gluteus maximus.

FUNCTIONAL ASPECTS

The movements at the hip joint

Movement at the hip joint can be considered either as a movement of the femur on the pelvis or of the pelvis on the femur. Both are important and as they occur under different conditions each will be considered in turn. The relation of the line of weight of the body above the hip joints to the hip joints (behind in the anatomical position) must always be considered when the body is in different postures.

The lower limb is in the **anatomical position** when it is in a straight line with the body and the big toe is pointing straight forwards. Flexion is the movement of the anterior surface of the thigh in the direction of the anterior abdominal wall. If the knee is extended, this movement is limited by tension of the hamstring muscles so that, on flexion of the knee, further flexion of the hip is permitted. Flexion is carried out by the iliopsoas assisted by the pectineus, rectus femoris and sartorius.

Extension beyond the anatomical position is limited to approximately 15°, the limitation being due to the large ligaments of the hip joint. Extension of the thigh is produced by the gluteus maximum and the hamstrings. Restoration of the flexed thigh to the anatomical position is due to gravity and is controlled by gradual relaxation of the flexors.

Abduction is the movement of the whole limb away from the midline and its range is about 40°. It is produced by the gluteus medius and minimus and is checked by the adductor muscles.

Adduction is movement of the limb towards and across the midline. It is produced by the adductor group and is limited by apposition of the two thighs, but can be much more extensive when the limbs are allowed to cross, or the other limb is abducted. From the abducted position, gravity adducts the thigh under the control of the abductors.

Lateral rotation occurs when the front of the femur rotates outwards round an axis passing from the centre of the head to the intercondylar fossa at the lower end of the femur. The range of this movement can be estimated by observing how far the foot is turned out with the knee extended. It is checked by the capsule and the range of movement is very variable. The muscles performing this movement are the obturators, piriformis, quadratus femoris and gluteus maximus.

Medial rotation is the opposite of lateral rotation and its range can be estimated by observing the inward movement of the foot. The range of movement is less than in lateral rotation as it is checked by the spirally directed fibres of the capsule and its thickenings. The muscles producing medial rotation are the anterior parts of the gluteus medius and minimus and the tensor fasciae latae.

Now consider the movements and muscles concerned when the femur is fixed or relatively fixed and the pelvis moves on it. Flexion, as defined before, occurs when the anterior surface of the thigh and anterior abdominal wall are approximated. This happens in bending the body forwards as in picking something off the ground. Restitution of the body to the upright position after this movement constitutes extension.

Flexion from the upright position is carried out by gravity, controlled by the orderly relaxation of the hamstrings and gluteus maximus muscles, but in the supine position, flexion of the trunk is carried out mainly by the iliopsoas. Conversely, extension from the stooping position is carried out by the hamstrings and gluteus maximus, acting at the hip joint. Extension from the sitting to the supine position is carried out by gravity controlled by the relaxation of the iliopsoas.

Abduction of the trunk on the lower limb is seen when one leg is lifted off the ground in the standing position and the whole weight of the body is carried by the supporting limb. When the right leg is raised, for example, the centre of gravity of the body is shifted towards the left side and the right side is prevented from dropping by the left gluteus medius and minimus (Fig. 35).

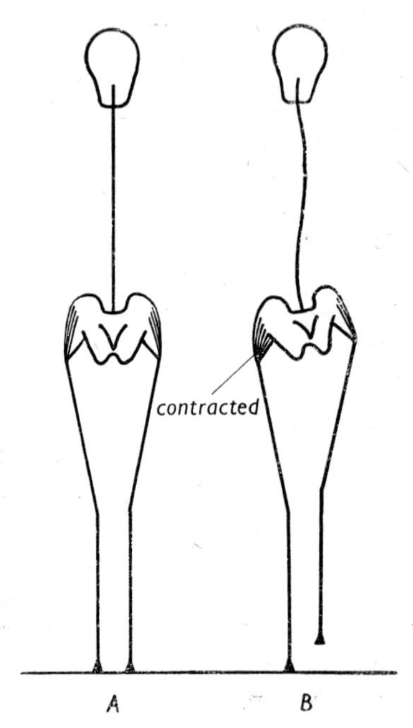

FIG. 35
In B the left gluteus medius is shown contracted. It abducts the pelvis on the left femur and, as a result, raises the right foot off the ground.

The use of the rotators of the hip is seen in every step of walking when the pelvis is medially rotated on the weight-bearing limb and the swinging limb is laterally rotated. This will be further considered in the section on walking in Chapter 22.

CHAPTER 18

THE KNEE JOINT

INTRODUCTION

THE knee is a hinge joint allowing chiefly flexion and extension. The whole weight of the body may be carried on the joint and balance is maintained by the action of muscle and ligament braces. The joint surfaces themselves are not congruous. Since no movement takes place laterally or medially the braces on the sides, as at all hinge joints, are ligamentous. In front, the joint is stabilised by a strong muscle brace and behind by both ligaments and muscles. The weight is transmitted wholly to the tibia.

The lower end of the femur is enlarged to form two **condyles** with two **epicondyles** as their most prominent medial and lateral areas. The linea aspera is continuous with two supracondylar lines, on the medial of which is the small **adductor tubercle** just above the condyle. The upper end of the tibia is also enlarged to form condyles. The **tibial tuberosity** is at the upper end of the anterior border of the tibia. On the under surface of the lateral condyle is the articular facet for the fibula. An oblique ridge (**soleal line**) passes downwards and medially across the upper part of the posterior surface of the tibia. The upper end of the fibula is enlarged to form the head and has an oblique articular facet for the tibia. The patella is a sesamoid bone which has a smooth articular posterior surface and a rough anterior surface. The upper edge is rounded and the lower edge is pointed. The lateral half of the articular surface is larger than the medial half.

DISSECTION

Continue the medial and lateral skin incisions in the thigh downwards to the ankle. Dissect the skin from the upper half of the leg in the form of anterior and posterior flaps. Remove the subcutaneous tissue and preserve on the medial side the great saphenous vein and saphenous nerve. Identify the **small saphenous vein** on the lateral side of the leg. It passes up the back of the calf and through the deep fascia at the level of the knee joint and joins the popliteal vein. Smaller veins will be found connecting the small with the great saphenous vein and perforating branches pierce the deep fascia to join the deep veins. Running with the distal part of the small saphenous vein is the **sural nerve,** which is formed by branches from the **tibial** and the **common peroneal nerves.** The posterior femoral cutaneous nerve lies in the superficial fascia and supplies the skin of the back of the thigh, knee and proximal half of the calf.

Remove the deep fascia on the medial side and the back of the knee joint and note that at the back it forms a roof for a diamond-shaped space, the **popliteal fossa.** On the lateral side preserve the iliotibial tract and trace it down to its attachment to the anterior surface of the lateral tibial condyle.

Remove the fat from the popliteal fossa. The sciatic nerve enters at its upper angle and divides into the tibial nerve medially and the common peroneal nerve laterally. Clean the muscles forming the boundaries of the fossa. The biceps forms the upper lateral boundary and the upper medial border is formed by the semimembranosus. The tendon of the semitendinosus lies on the posterior surface of the semimembranosus. The inferior boundaries of the popliteal fossa are formed laterally by the lateral head of the **gastrocnemius** and the **plantaris** and medially by the medial head of the gastrocnemius. Follow the latter muscles upwards to their origins from the back of the femur. The femoral artery (deep) and vein (superficial) pass downwards in the fossa as the **popliteal artery** and **vein,** deep to the tibial nerve, and by separating the heads of the gastrocnemius the artery can be seen to divide into the anterior and posterior tibial arteries. Articular and muscular branches are given off in the fossa.

At the lower end of the fossa, dissect deep to the artery and clean the surface of the **popliteus muscle.** It will be more fully exposed later.

Clean the tendons lying on the medial side of the joint and in front of the semimembranosus. These are the tendons of the sartorius and gracilis ; trace them, and also the tendon of the semitendinosus, to their attachments to the medial side of the upper end of the tibial body. The posterior muscles brace the knee at its back and flex the tibia on the femur.

The upper attachment of the flexors of the knee (hamstrings) to the ischial tuberosity has already been noted. The lateral part of the hamstring muscles is formed by the **biceps femoris,** which, in addition to its attachment to the ischial tuberosity (long head), is attached to the back of the femur (short head). The two portions join and end in a common tendon, which is attached to the head of the fibula.

The **semitendinosus** and **adductor gracilis** are attached to the upper part of the medial surface of the tibia where the sartorius is also attached. The **semimembranosus** is attached to the posterior part of the medial condyle of the tibia and sends fibres to the back of the capsule of the knee joint. The **gastrocnemius** is attached by lateral and medial heads to the back of the condyles of the femur. The **plantaris** is attached by a small muscular belly to the lateral condyle (Fig. 37).

If the four components of the **quadriceps muscle** have not been clearly displayed, they must now be cleaned. The rectus femoris has already been identified and the other parts of the quadriceps muscle are related to it (Figs. 32, 33, 36). The **vastus medialis** is attached to the body of the femur and the linea aspera. It remains muscular to the level of the patella. The **vastus lateralis** is attached to the linea aspera and extends upwards as far as the greater trochanter, passing forwards to the intertrochanteric line. It becomes aponeurotic above the level of the patella. The **vastus intermedius** lies deep to the rectus femoris and vastus lateralis and is attached to the body of the femur.

Examine the whole muscle, particularly its lower attachment. It is inserted as a tendon within which the patella lies forming a sesamoid bone. This tendon passes as a relatively narrow **patellar ligament** to its insertion on the tibial tuberosity (Fig. 36). These four large muscles all pull on this point and thus extend the leg at the knee joint. Note that part of the quadriceps insertion is aponeurotic, forming thin but strong sheets attached to the sides of the patella and the tibial condyles, and that these are continuous laterally and medially with the joint capsule. In other words, these aponeurotic expansions from the quadriceps (**patellar retinacula**) form part of the capsule of the joint.

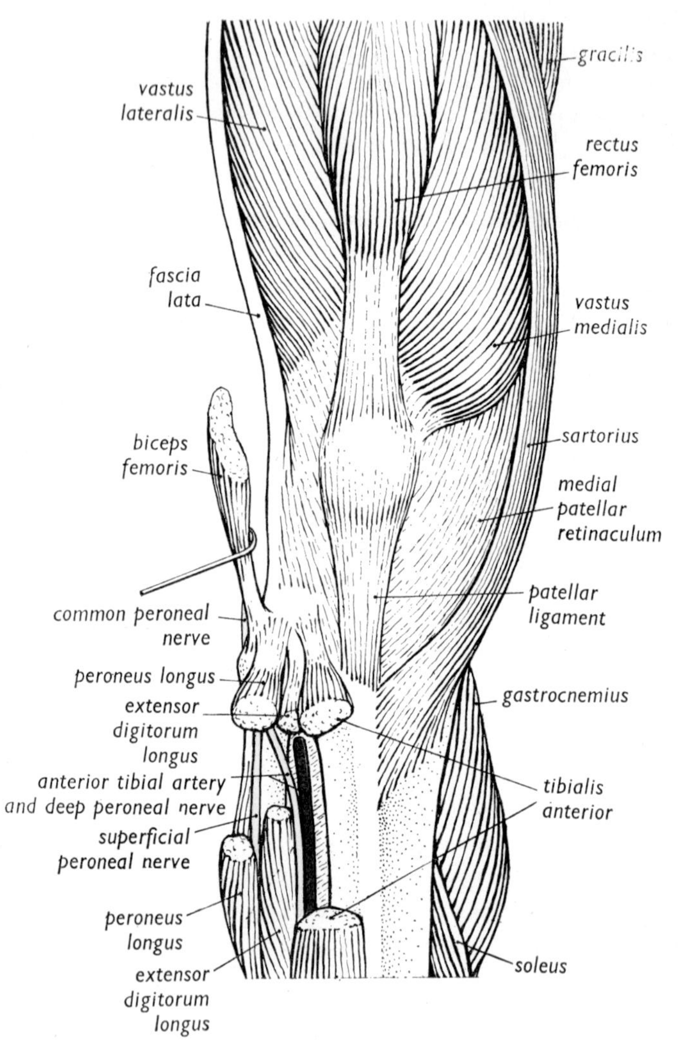

FIG. 36
Diagram of the anterior aspect of the knee joint.

Cut through all the structures in the thigh by a circular incision down to the bone at a level 10 cm above the upper margin of the patella. Pull away all tendons except the quadriceps femoris from the neighbourhood of the joint so as to get good access to it. Remove the two heads of the gastrocnemius and the plantaris from the femur

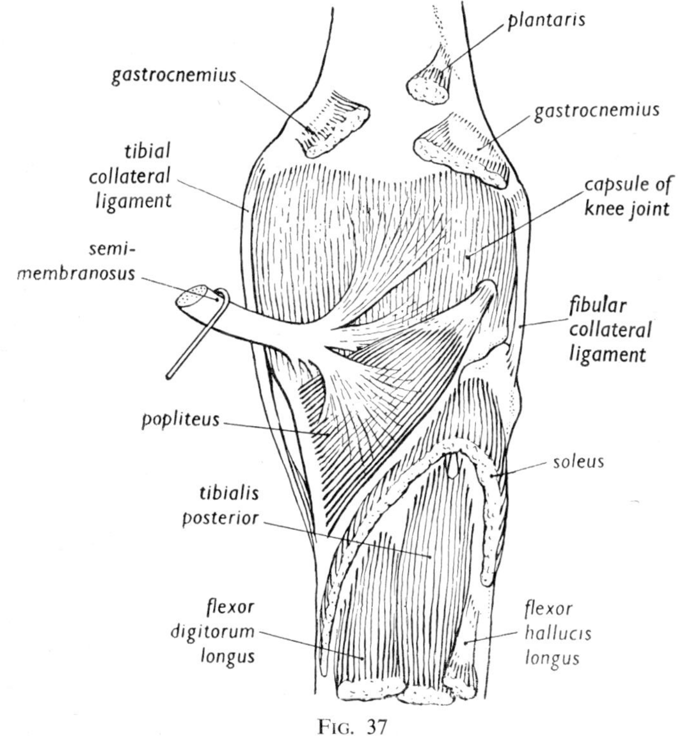

FIG. 37
Diagram of the posterior aspect of the right knee joint.

and turn them downwards. Take this opportunity to examine carefully all the muscle attachments round the joint. Note that the muscles fall into anterior (extensor) and posterior (flexor) groups, which together balance the thigh on the leg.

Lift the quadriceps femoris carefully off the front of the femur and as the muscle is raised look for a thin layer of synovial membrane passing from the front of the femur to the back of the quadriceps. This is the upper boundary of the **suprapatellar bursa** of the synovial cavity (Fig. 38). Its upper edge is usually 6 cm above the upper border of the patella when the knee is extended. On the sides and posterior surface of the joint, clear away any loose connective tissue that may be obscuring the view of the capsule. On the

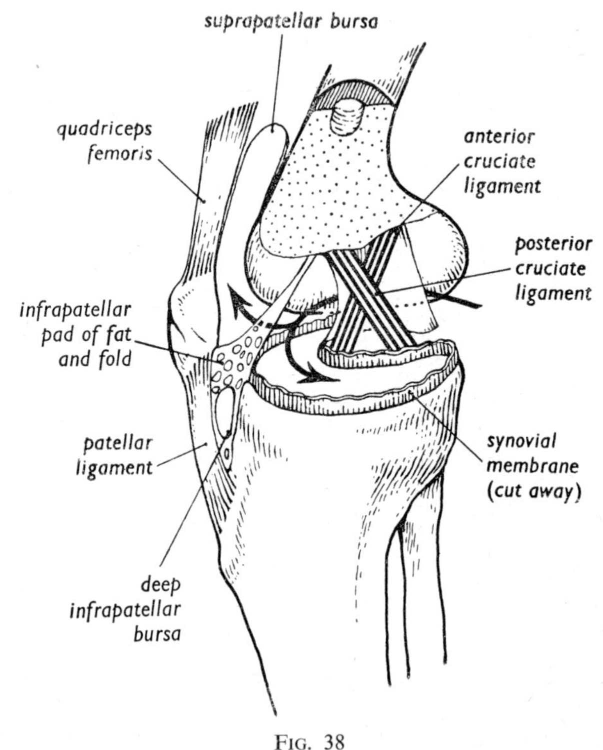

FIG. 38

Diagram of some of the synovial relations in the knee joint. The two cruciate ligaments are in the intercondylar septum. The arrow indicates the communication between the condylar pouches and the suprapatellar bursa.

lateral side, expose and clean the **fibular collateral ligament** of the joint, running from the lateral epicondyle to the upper end of the fibula. Note that it is not attached to the capsule.

After cleaning the surface of the capsule, examine its attachments and identify on the back, the **oblique popliteal ligament** running upwards and laterally from the medial tibial condyle as an extension of the fibres of the semimembranosus (Fig. 37). Medially identify a strong and important capsular thickening, the **tibial collateral ligament,** running from the medial epicondyle to the medial tibial condyle. Note that it is attached to the margin of the tibial condyle and extends down the medial side of the upper end of the tibia. Finally trace the tendon of the popliteus upwards and laterally and verify that it pierces the back of the capsule just below the lateral femoral condyle.

Make vertical incisions in the retinacula on both sides, 4 cm behind the patella and **gently** pull the quadriceps and patella downwards and forwards. Observe that between the upper surface of the tibia and the level of the lower border of the patella two synovial fringes, the **alar folds,** project into the joint on either side. They contain fat which is continuous with the infra-patellar pad of fat lying between the patellar ligament and tibia. The folds meet in the middle of the joint, forming a single **infrapatellar synovial fold,** which is attached to the most anterior edge of the intercondylar fossa of the femur (Fig. 38). (*N.B.*—This fold may be torn if the patella is pulled down roughly.) Behind this fold will be seen the **anterior cruciate ligament,** which with the **posterior cruciate ligament** invaginates the synovial membrane from behind. On the upper surface of the tibia identify the anterior horns of the **medial** and **lateral menisci** (semilunar cartilages).

Cut the infrapatellar fold and then cut through the capsule all round the joint close to its femoral attachments, keeping well above the menisci. Also detach the fibular collateral ligament from the lateral epicondyle and the tendon of popliteus from the lateral condyle of the femur. Examine the anterior and posterior cruciate ligaments. Viewing the joint from the lateral and medial side in turn, flex and extend the **tibia on the femur.** During flexion the menisci and the tibial head come in contact with the more posterior parts of the femoral condyles. Towards the end of extension the cruciate ligaments tighten. The anterior ligament stops the rolling movement between the lateral condyles but the medial tibial condyle continues to glide on the femoral condyle for a short distance and this is accompanied by a lateral rotation of the tibia on the femur. The position of full extension can be maintained with the minimum of muscular activity, because the line of weight falls in front of the knee joint when standing erect. The popliteus medially rotates the tibia on the femur at the beginning of flexion.

Flexion of the **femur on the tibia** is preceded by lateral rotation of the femur due to the popliteus, and at the end of extension of the femur on the tibia, the femur rotates medially, probably due to the tightening of the ligaments and the differences in shape and size of the condyles. Also note that the

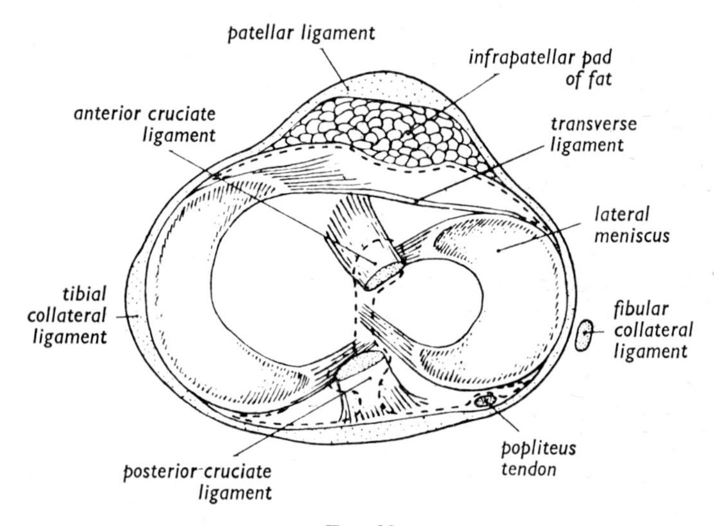

FIG. 39

Diagram of the structures seen on the upper surface of the right tibia. The line of attachment of the synovial membrane is shown by the broken line.

patella moves downwards on the femur during flexion. Cut through the attachment of the anterior cruciate ligament to the lateral femoral condyle and of the posterior cruciate ligament to the medial femoral condyle and separate the tibia from the femur.

Examine the two menisci on the upper surface of the tibia and note that the medial is less circular than the lateral (Fig. 39). Confirm that their peripheral surfaces are attached to the inner surface of the capsule and to the tibia. Try to determine which of the two is more firmly attached to the capsule. It should be the medial meniscus, but it may be difficult to demonstrate this on a preserved specimen. Examine the attachments of the cruciate ligaments to the tibia and note that the anterior ligament is attached to the tibia between the anterior ends of the menisci and that the posterior is attached to the tibia behind their posterior ends. Finally look at the lower end of the femur and define the lines between the patellar and tibial articular surfaces on the lateral and medial condyles. These are more easily seen on a fresh specimen than on a dried bone.

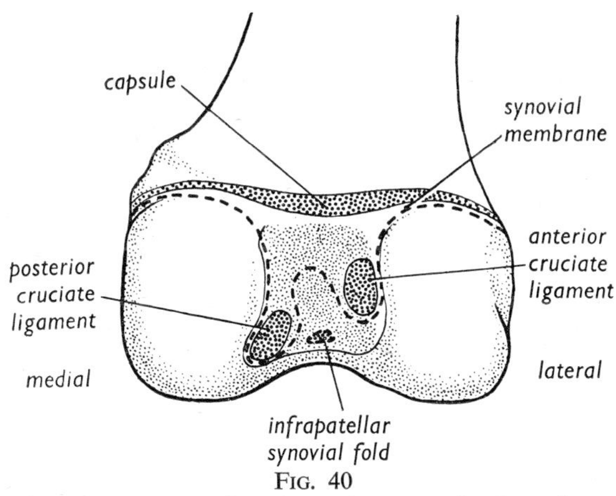

FIG. 40

Diagram of the relations in the intercondylar fossa from behind. The reflexion of the synovial membrane is shown by the broken line.

STRUCTURAL DETAILS

The femur

The linea aspera is a well-developed ridge on the middle third of the posterior surface of the body. The rest of the middle of the shaft is rounded and smooth and a transverse section at this level is pear-shaped with the apex posteriorly. The linea aspera (Fig. 29) divides distally into two ridges, the medial and lateral supracondylar lines. The medial line leads to the adductor tubercle, immediately proximal to the medial condyle. The condyle is divided into a smooth articular surface and a roughened medially placed surface on which the most prominent bony area is the **medial epicondyle.** The lateral supracondylar line runs down to the lateral condyle, which has a smooth articular surface and a roughened lateral surface on which is the bony prominence of the **lateral epicondyle.** Below the lateral epicondyle is a groove for the popliteus tendon. Between the two supracondylar lines is the smooth, triangular, popliteal surface, continuous distally with the **intercondylar fossa,** which lies posteriorly between the two condyles (Fig. 40).

The anterior surface of the femur is smooth and rounded, and broadens out distally above the two condyles.

Looked at from below, the articular surfaces are seen to be separated by the intercondylar fossa posteriorly. Anteriorly they unite in a concave patellar articular surface. An ill-defined groove across the articular portion of each condyle separates the patellar from the tibial articular surfaces. The medial tibial articular surface is seen to be slightly longer from back to front than the lateral.

When placed in the anatomical position the inferior surfaces of the condyles of the femur are in a horizontal plane. The body makes an angle of about 10° with the vertical. This angle is greater in the female than in the male. There is also an anterior convexity of the body.

The patella

The patella has a small projection, the apex, pointing distally. The anterior surface is rough and the posterior surface is smooth for articulation with the femur except for a narrow margin superiorly and the apex below. The articular surface has a vertical ridge medial to its midline, so that the medial facet is smaller than the lateral.

The tibia

The tibia, although triangular in transverse section in its middle third, is, like the femur, a bony tubular girder. The upper end is much larger than the lower. There is a very prominent anterior border with the **tibial tuberosity** at its upper end for the attachment of the patellar ligament. The expanded head has lateral and medial condyles for articulation with the femur, the medial being larger. The two articular surfaces are separated by the rough intercondylar area.

Viewed from the front, the tuberosity is continuous above with the margins of the medial and lateral condyles, and narrows below to an apex continuous with the anterior border of the body. Posteriorly the body has a smooth surface, crossed by the oblique **soleal line,** which runs downwards and medially. On the postero-inferior surface of the lateral condyle is a small oblique facet for the fibula.

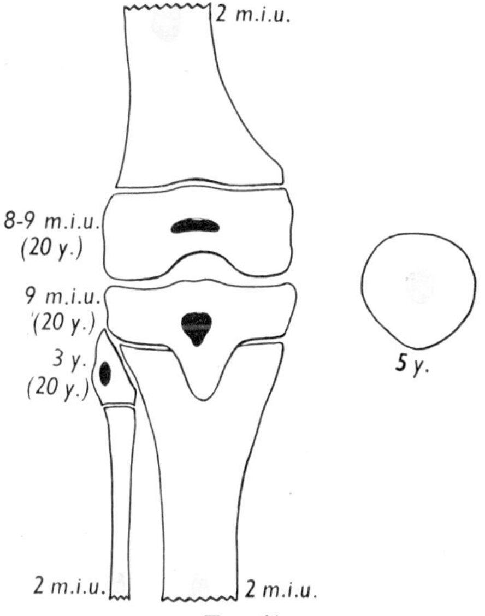

FIG. 41

The primary centres of ossification of the bones forming the knee joint are shown in yellow and the secondary centres in red. The figures in brackets indicate the times of fusion of the primary and secondary centres (y., years; m.i.u., months in utero).

The fibula

The upper end **(head)** of the fibula has an obliquely placed articular facet on its anterosuperior aspect, and above and posterior to the articular facet is the **apex.** In order to be able to determine to which side a fibula belongs, examine its ends and find the **malleolar fossa** which is distal, medial and posterior. When the tibia and fibula are articulated the head of the fibula is below the level of the tibial condyle and the upper end of the body of the fibula is posterior as well as lateral to the tibia.

Ossification (Fig. 41)

Primary centres appear in the bodies of the tibia and the fibula at about the end of the second month of intra-uterine life. The lower femoral epiphysis begins to ossify before the ninth month of pregnancy and the upper tibial epiphysis follows soon after. Sometimes there is a separate ossification centre for the tibial tuberosity at about the tenth year. The head of the fibula has a separate centre appearing at about the third year. All these centres join the bodies at about twenty years. Ossification in the patella begins at about the fifth year and is completed by puberty.

The capsule of the knee joint

The main movements at the joint are flexion and extension. The capsule is therefore thickened and taut at the sides. Its femoral attachment is to the popliteal surface just above the condyles and to the upper margin of the intercondylar fossa (Fig. 40). Traced forwards on either side, it is attached just below the epicondyles. In front, the capsule is replaced by strong expansions from the tendon of the quadriceps femoris muscle, the **patellar retinacula,** which are attached to the sides of the patella and are continuous with the capsule posteriorly. In the midline, the quadriceps tendon, the patella and the patellar ligament take the place of the capsule. Below, the capsule is attached to the condyles of the tibia just distal to the edge of the articular surfaces except anteriorly. Here, as the patellar ligament forms the central portion of the capsule, the line of attachment runs in a V-shape from the anterior surface of the tibial condyles towards the tibial tuberosity (Fig. 36). The menisci are attached at their periphery to the inner surface of the capsule and inferiorly to the tibia.

The ligaments (Figs. 36 and 37)

The **tibial collateral ligament** is a very strong broad capsular thickening attached above to the medial epicondyle and below to the medial side of the medial tibial condyle and the upper end of the body. Its function is to prevent abduction at the knee. The **fibular collateral ligament** is a strong, rounded cord passing from the lateral epicondyle to the apex of the fibula. It is separate from the capsule and prevents adduction at the knee.

The **oblique popliteal ligament** is a capsular thickening running from the back of the medial tibial condyle to the back of the lateral femoral condyle and is an upward and lateral expansion of the attachment of the semimembranosus.

The intracapsular structures (Figs. 38, 39 and 40)

There are two main intracapsular ligaments, the **anterior** and **posterior cruciate.** Their names, anterior and posterior, refer to their tibial attachments, the anterior being attached in front and the posterior well behind in the intercondylar area. The anterior is attached above to the medial side of the lateral femoral condyle, and the posterior to the lateral side of the medial femoral condyle. They are strong round ligaments and their function is to check sliding in an anteroposterior direction between the tibia and the femur. The anterior ligament prevents the tibia from sliding too far forwards on the femur and the posterior ligament stops it sliding too far back. The anterior ligament becomes taut in extension. These ligaments also limit rotatory movements of one bone on the other.

Besides the cruciate ligaments, there are also within the capsule the tendon of the popliteus muscle and the lateral and medial menisci. The popliteus tendon arises from a small depression below the lateral epicondyle, where it lies within the capsule. It passes posteriorly and downwards, grooving the posterolateral edge of the lateral meniscus, and then passes out of the capsule through a hole just behind the lateral tibial condyle (Fig. 37). The menisci of the joint are two crescentic pieces of fibrocartilage, thick at their periphery and thin at their inner edge. They rest on the upper surface of the tibial condyles and are attached to the condyles by the innermost fibres of the capsule, except where the ends of the menisci leave the capsule and pass towards the intercondylar area (Fig. 39). The lateral cartilage is the smaller, forming about three-fifths of a circle, and its ends or horns are embraced by the two horns of the medial cartilage which forms a crescent of larger diameter. In addition to the capsular attachment, each cartilage is attached by its anterior and posterior horns to the upper surface of the tibia between the articular areas. The anterior ends of the cartilages are also united by the transverse ligament. The popliteus tendon is attached to the back of the lateral meniscus and separates it from the capsule.

The part played by the cartilages of the joint is uncertain. Like other articular discs, they probably serve to improve the congruity between the bony surfaces. As they are attached to the tibia, they move with this bone. After operative removal in man, regeneration takes place, not of the cartilage, but of a wedge-shaped mass of fibrous tissue occupying the same position and of about the same size. In man, the removal of a meniscus does not usually interfere with the efficient working of the joint.

The synovial membrane (Figs. 38 and 39)

The synovial membrane follows approximately the rule for all such membranes, that is, it lines the internal surface of the non-articular structures. But there are certain places in the joint where this rule does not hold. It is therefore helpful to consider the development of the joint. Early in fetal life, instead of one synovial cavity there are three. One lies between the patella and the front of the femur, and one between each femoral and tibial condyle. The partition separating the patellar cavity from the intercondylar cavities breaks down on each side, so that all three cavities are in communication with each other. The infrapatellar and alar folds, however, are left as vestiges of this partition, and part of the partition between the two intercondylar cavities remains as the intercondylar septum. In it are the two cruciate ligaments.

Anteriorly, the membrane passes backwards as a thin fold in the midline of the joint from the lower edge of the patella to be attached to the anterior end of the femoral intercondylar fossa. This forms the **infrapatellar fold,** a flattened cone of synovial membrane with its apex attached to the femur and its base stretching between the inferior border of the patella and the anterior end of the intercondylar area of the tibia. Inside this cone is the infrapatellar pad of fat (Fig. 38). As the infrapatellar fold passes backwards it carries with it on either side two horizontal folds continuous with the infrapatellar fold but having a free edge. These alar folds lie between the femoral and tibial condyles.

Posteriorly, where the synovial membrane does not line the capsule in the midline of the joint, it forms the **intercondylar septum,** lying in front of and on either side of the cruciate ligaments and dividing the joint cavity, as seen from the back, into two pouches, one under each femoral condyle.

The synovial membrane passes from the upper border of the patella along the deep surface of the quadriceps femoris tendon to a point about 6 cm above the patella. It then turns downwards on the anterior surface of the femur as far as the margin of the articular surface. This pouch of synovial membrane, lying above the patella, in front of the femur and behind the quadriceps muscle is known as the **suprapatellar bursa.**

Bursae

There are many bursae associated with the knee joint. The most important are : the suprapatellar (described above), the prepatellar (in front of the patella), the two infrapatellar (subcutaneous and deep to the patellar ligament) and the one between the semimembranosus tendon and the tibia. The suprapatellar and the semimembranosus bursae communicate with the joint. There are also bursae associated with some of the other tendons and ligaments round the joint. The prepatellar bursa may be injured in certain occupations.

The quadriceps femoris

Three parts of the muscle arise from the front and sides of the femur and the fourth, placed a little in front of the others, arises by two heads, one from the anterior inferior spine of the ilium and the other from the upper margin of the acetabulum. The first three muscles are the vastus intermedius, vastus lateralis and vastus medialis and the fourth is the rectus femoris. These four muscles join and are inserted by means of the patella and the patellar ligament into the tuberosity of the tibia. The vasti cover the curved surfaces of the body of the femur, leaving only the linea aspera for the attachment of other muscles. Vastus lateralis has a tendinous origin from the lateral edge of the linea aspera, and continues upwards lateral to the gluteal tuberosity as high as the greater trochanter. Vastus intermedius is attached to the lateral and anterior surfaces of the upper two-thirds of the bone Vastus medialis is attached to the medial edge of the linea aspera and covers, but does not take attachment from, the medial side of the femur.

All four muscles are attached below to the patella. The rectus femoris and vastus intermedius go to the upper border, with the rectus in front of the intermedius, the vastus lateralis to the lateral side and the vastus medialis to the medial side of the patella. From the patella, the **patellar ligament** runs to the inferior part of the tuberosity of the tibia. Through the patellar retinacula, the vastus lateralis and the vastus medialis are attached to the tibial condyles.

The popliteus

This muscle is intracapsular where it is attached above to the lateral condyle of the femur. The muscle pierces the capsule, and its lower attachment is to the posterior surface of the tibia above the oblique soleal line. When the muscle contracts, it medially rotates the tibia on the femur, and pulls the posterior end of the lateral meniscus backwards.

Vessels and nerves

The **popliteal artery,** the continuation of the femoral artery, begins at the opening in the adductor magnus, through which the artery passes. From this point it runs vertically downwards to the lower border of the popliteus, where it divides into the anterior and posterior tibial arteries. The popliteal artery supplies branches to the muscles, to the skin and to the knee joint. The **popliteal vein** lies superficial to the artery in the fossa. Its largest tributary is the small saphenous vein which joins it by piercing the deep fascia roofing over the popliteal fossa.

The **sciatic nerve** enters the popliteal fossa at its apex and usually divides there into tibial and common peroneal nerves, though in a small number of cases this division takes place in the gluteal region. The **common peroneal nerve** passes downwards and laterally along the medial border of the biceps muscle (Fig. 27). It passes behind the head of the fibula and leaves the popliteal fossa by turning forwards and downwards lateral to the neck of the fibula. It gives off cutaneous and articular branches in the fossa, including the lateral sural cutaneous nerve. The **tibial nerve** runs downwards crossing superficial to the artery and vein from the lateral to the medial side. In the upper part of the fossa it gives off cutaneous branches, which go to the back and lateral side of the lower part of the leg; lower down it gives off branches to some of the muscles of the calf (gastrocnemius, soleus, plantaris), to the popliteus and to the knee joint.

FUNCTIONAL ASPECTS

The movements at the knee joint

At the knee joint the main movements are flexion and extension, and may be produced with either the tibia or the femur fixed. In many movements, such as walking, both bones move at the same time (Chapter 22).

The anatomical position of the joint is that in which the femur and tibia form a straight line when viewed from the lateral side. In this position the greater trochanter, the lateral epicondyle and the lateral malleolus are in the same coronal plane. Flexion occurs when the calf of the leg is approximated to the posterior surface of the thigh, and extension is the return of the leg to the anatomical position. There is possible, in most limbs, a further movement of about 5° of **hyperextension** beyond the anatomical position.

In the anatomical position, as defined above, the lower part of the patellar articular surface is in contact with the upper portion of the articular surface on the femur. This high position of the patella is seen in a lateral radiograph of the knee. As flexion proceeds the patella glides down the articular surface on the femur and in full flexion only the upper and medial part of the patella is in contact with the femur. In the anatomical position, the anterior portions of the femoral condyles are in contact with the tibial condyles, particularly their anterior parts, which take the main thrust of the body weight. The position of the tibia relative to the femur changes during flexion, so that in extreme flexion the posterior portions of the femoral condyles remain in contact with the anterior part of the tibial condyles. It will be seen, therefore, that the femoral and tibial surfaces slide and roll upon each other. The functions of the menisci in these changes are uncertain. They move with the tibia.

During the last part of extension of the tibia on the femur the tibia rotates laterally to a small extent on the femur, and conversely when the tibia is fixed, the femur at the end of extension rotates medially on the tibia, producing what is known as the " locking " of the joint. This is produced partly by the medial condyle being longer anteroposteriorly than the lateral condyle, and partly by the arrangement of the cruciate ligaments. Extension of the joint is produced by the quadriceps femoris. The importance of the quadriceps muscle as a brace for the maintenance of balance at the joint cannot be overstressed. When full extension has been reached and the knee is " locked ", the quadriceps relaxes. This can be confirmed by noting the mobility of the patella of a person standing with the knees fully extended. Flexion is produced by the hamstring muscles and gastrocnemius, and is limited by the apposition of the surfaces of the calf and thigh. Medial rotation of the tibia on the femur at the beginning of flexion is produced by the popliteus.

Rotation of the tibia on the femur, or vice versa, apart from the rotation at the beginning of flexion and at the end of extension, is limited by the obliquity of both the collateral ligaments, and by the cruciate ligaments. A small amount of active rotation is possible when the knee is semiflexed. Forward and backward gliding of the femur and tibia on each other is prevented by the cruciate ligaments, although a small amount of passive movement of this type can be demonstrated if the knee is semiflexed.

When the quadriceps femoris contracts it tends to pull the patella laterally as well as upwards. This lateral movement is checked by the anterior prominence of the lateral condyle and by the lowest horizontal fibres of the vastus medialis which are inserted on the medial side of the patella.

CHAPTER 19

THE LEG AND THE DORSUM OF THE FOOT

INTRODUCTION

IN this section, the principal structures to be examined are the muscles that act on the ankle joint and on some of the joints of the foot. The deep fascia of the leg (fascia cruris) is thick and fibrous. It is continuous with the fascia lata above and forms a close investment for the leg. Many muscles are attached to its deep surface. Prolongations of the fascia cruris form septa between the muscles, giving additional surfaces for the attachment of muscle fibres, and divide the leg into anterior, posterior and lateral compartments (Fig. 42). The anterior intermuscular septum passes to the anterior border of the fibula and separates the anterior (extensor) compartment from the lateral (peroneal) compartment. The pos-

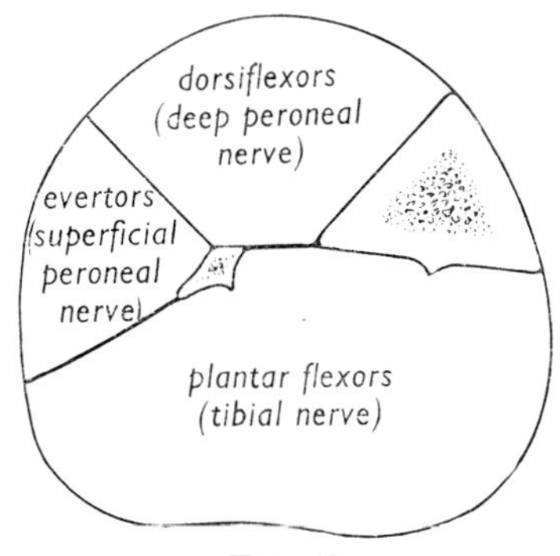

FIG. 42
Diagram of a section of the middle third of the leg showing the fascial septa and the muscle compartments. The fibula is on the left and the tibia, with its subcutaneous surface, is on the right.

terior intermuscular septum passes to the posterior border of the fibula and separates the lateral (peroneal) compartment from the posterior (flexor) compartment. Medially, the deep fascia is continuous with the periosteum on the subcutaneous surface of the tibia.

Examine the skeleton and note that most of the fibula is in a plane posterior to that of the tibia. At the ankle, the lower ends of the bones have projecting processes called **malleoli** forming the sides of the mortise into which the **talus** fits. The heel is formed by the backward projecting mass of the **calcaneus** which articulates with the talus above and the **cuboid** in front. Between the talus and the calcaneus is a tunnel called the **sinus tarsi.** The talus articulates in front with the **navicular** which articulates with the three **cuneiforms.** The three cuneiforms and the cuboid articulate with the five **metatarsal bones.** The **phalanges** of the toes are similar to those found in the hand though the middle phalanges are small and in the fifth toe the terminal two phalanges may be fused.

The tibia transmits all the weight from the femur to the talus, and is of robust construction. The fibula, on the other hand, serves mainly for muscle attachments and its lower end forms the lateral part of the mortise for the talus. The weight is distributed from the tibia to the ground partly backwards to the heel and partly forwards to the front of the foot. The bones of the tarsus are modified from the basic mammalian pattern as are the hand bones (page 68). There are only two bones in the proximal row, the talus and calcaneus. The talus is the only bone articulating with the tibia and it forms the keystone of the longitudinal arch of the foot. It rests on the calcaneus and articulates with the navicular in front. Weight passes from the navicular to the cuneiforms and the medial three metatarsals. Some weight is transmitted to the cuboid from the calcaneus and from the cuboid along the two lateral metatarsals.

The ankle joint is a hinge joint at which the weight of the body is balanced, and active movements of **plantar flexion** (flexion) and **dorsiflexion** (extension) occur. Muscle braces are therefore provided at the front and back but the sides of the joint are stabilised by strong ligaments and the malleoli. Many of the muscles to be dissected act on the tarsal joints distal to the ankle. Some muscles turn the sole of the foot inwards and adduct the medial border **(inversion)** and others turn it outwards and abduct the lateral border **(eversion).**

DISSECTION

The front of the leg (Figs. 36 and 43)

Extend the skin incisions of the leg along the lateral and medial sides of the foot. Turn down the anterior flap to the ends of the toes and the posterior flap as far as the heel.

Trace the **saphenous nerve** and **great saphenous vein** down as far as the lower end of the tibia and remove the subcutaneous fascia from the front of the leg. Examine the deep fascia and observe that it forms two strengthening bands across the ankle (the **extensor retinacula**). The upper binds down the tendons above the ankle and the lower binds them down in front of the joint (Fig. 45).

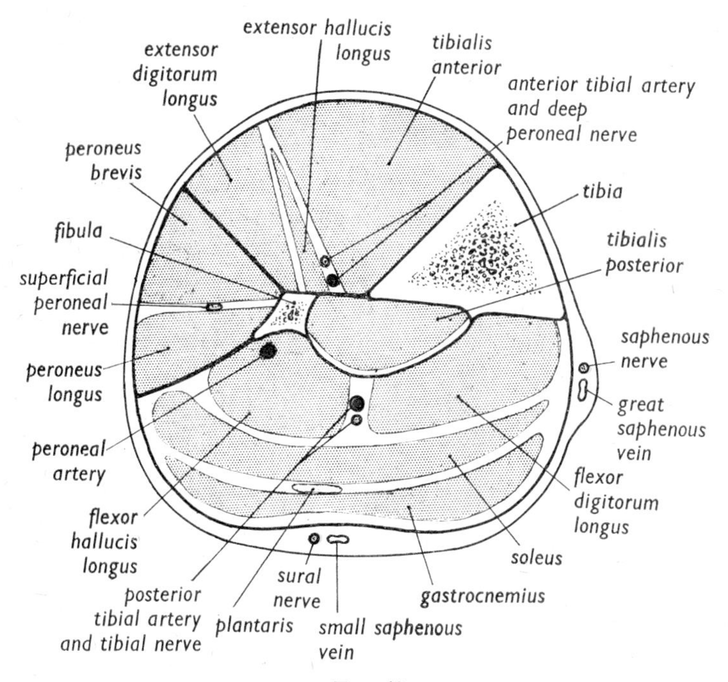

FIG. 43
The principal structures seen in a transverse section of the middle third of the leg.

Cut through the deep fascia, remove it from the front of the leg and the dorsum of the foot but preserve the retinacula. As the fascia is removed, note that some muscle fibres arise from it. On the dorsum of the foot there is a subcutaneous venous plexus.

In front of the ankle joint identify the following tendons from the medial to the lateral side, the tibialis anterior, extensor hallucis longus and extensor digitorum longus. All of these muscles brace the front of the joint. Pick up each tendon at the ankle joint and trace it upwards to its muscular belly. Between the tibialis anterior and the extensor muscles in the leg find the anterior tibial artery and deep peroneal nerve. Follow them downwards to the ankle joint where the extensor hallucis longus is medial to the vessels and nerve, and upwards as far as the level of the tibial tuberosity; note that in the leg they lie on the interosseous membrane. Examine the extensor digitorum longus, the lower part of which may be separated off and form another muscle, the peroneus tertius.

The **extensor digitorum longus** and **extensor hallucis longus** are attached to the anterior surface of the fibula. In the lower third the extensor digitorum longus is usually continued as the **peroneus tertius.** The muscles are also attached to the adjacent interosseous membrane and to the fascia covering them. The **tibialis anterior** is attached to the lateral surface of the tibia, the adjacent interosseous membrane and the covering fascia (Figs. 36 and 43).

Trace the extensor tendons downwards, along the dorsum of the foot to their distal attachments, the extensor hallucis longus to the base of the distal phalanx of the big toe and the extensor digitorum longus into a dorsal expansion on the back of the proximal phalanx of the lateral four toes. This dorsal expansion is attached to the middle and distal phalanges. The tibialis anterior is attached distally to the medial surface of the medial cuneiform and the base of the 1st metatarsal bone. The peroneus tertius is attached to the dorsal surface of the 5th metatarsal.

All these muscles are dorsiflexors (extensors) of the ankle joint and the extensor digitorum longus and extensor hallucis longus also dorsiflex the toes. The tibialis anterior inverts (turns the sole inwards) and the peroneus tertius everts the foot (turns the sole outwards). They are all supplied by the deep peroneal nerve. Follow the anterior tibial artery on to the dorsum of the foot where it is known as the **dorsalis pedis artery,** and the terminal branch of the deep peroneal nerve to the web between the great and second toes.

On the dorsum of the foot a small muscle, the **extensor digitorum brevis,** is found. It is attached proximally to the calcaneus in front of the fibula. Trace its four tendons to their distal attachments. The most medial tendon (extensor hallucis brevis) is attached to the base of the proximal phalanx of the great toe and the three lateral tendons join the medial three tendons of the extensor digitorum longus.

The lateral aspect of the leg

Examine the muscular belly of the extensor digitorum longus. On its lateral side there is a fascial septum passing deeply to be attached to the fibula. This septum is the anterior boundary of the **peroneal compartment** of the leg, which lies lateral to the fibula. It contains two muscles, the peroneus longus (above) and the peroneus brevis (below), and the branches of the common peroneal nerve (Fig. 36). Note that the common peroneal nerve as it lies on the neck of the fibula is nearly subcutaneous with very little of the peroneus longus separating it from the skin. Here it can easily be rolled on the bone. Because of this position it may be involved in fractures of the neck of the fibula or affected by external pressure. Trace the nerve from where it leaves the popliteal fossa, through the substance of the peroneus longus and round the neck of the fibula. The nerve divides into the deep peroneal nerve, already seen in the anterior compartment, and the superficial peroneal nerve passing downwards between the two peronei muscles and supplying them. The latter nerve becomes subcutaneous at the junction of the middle and lower thirds of the leg. After tracing this nerve and its branches over the dorsum of the foot, clean the peroneal muscles and dissect out their course on the side of the foot. The brevis passes to the base of the fifth metatarsal, and the longus passes round the lateral side of the foot into the sole (Fig. 58). On the lateral surface of a calcaneus identify the peroneal trochlea. Find this pulley on the dissection and ascertain the relation of the peroneal tendons to it (brevis above and longus below).

The **peroneus longus** and **brevis** are attached to the lateral surface of the fibula. As they pass behind the lateral malleolus the tendon of the peroneus brevis is anterior to that of the peroneus longus. The peroneus longus and brevis are evertors of the foot and are supplied by the superficial peroneal nerve.

The back of the leg (Figs. 37 and 43)

Remove the superficial fascia from the back of the leg, preserving the great and small saphenous veins ; with the former is the saphenous nerve and with the latter is the sural nerve, a branch of the tibial nerve. Cut through the deep fascia by a median longitudinal incision and remove it from the surface of the muscles lying deep to it. It should be removed from the whole of the back and medial side of the leg to beyond the medial malleolus and from the back of the heel.

The gastrocnemius muscle, whose origin was examined in the dissection of the knee joint, is exposed. Identify its two bellies passing into the single **tendo calcaneus,** a large and powerful tendon attached to the back of the calcaneus (Fig. 57). Trace the long thin tendon of the plantaris muscle from its attachment on the back of the lateral femoral condyle, and then separate the bellies of the gastrocnemius muscle from the structures deep to them. Make a vertical cut between the fused heads of the gastrocnemius and expose the **soleus muscle.** Examine its attachments to the upper parts of the tibia and fibula and note that its tendon blends with that of the gastrocnemius. These muscles are braces of the back of the ankle joint and plantar flex the foot on the leg. Note that there are large veins in the substance of soleus and between the two muscles.

Cut through the tendo calcaneus and turn the gastrocnemius and soleus muscles upwards towards the knee. The aponeurotic attachment of the soleus to the soleal line of the tibia should be carefully incised. The deep structures of the leg are now exposed. Clean the posterior tibial artery and the tibial nerve from the popliteal region to the level of the medial malleolus. Find the peroneal branch of the artery laterally and the branches from the nerve to the muscles of the calf. Examine the deep muscles. At the upper end of the calf note the attachment of the **popliteus** to the back of the tibia above the soleal line. Distal to the popliteus are the **flexor hallucis longus** arising from the back of the fibula, the **flexor digitorum longus** arising from the back of the tibia medially and, placed more deeply than these two and arising from both bones, the **tibialis posterior.** Follow all three tendons downwards to the back of the medial malleolus and note how they are arranged there in relation to each other and to the artery and nerve. From the medial to the lateral side are the tibialis posterior, the flexor digitorum longus, the posterior tibial artery and veins, the tibial nerve and the flexor hallucis longus. These muscles assist gastrocnemius and soleus as braces of the back of the ankle and also have actions on the joints of the foot distal to the ankle joint.

Find the origin of the **anterior tibial artery** from the popliteal artery at the lower border of the popliteus muscle and trace it forwards through the interosseous membrane between the tibia and fibula into the anterior compartment of the leg.

The **flexor digitorum longus** is attached to the medial part of the posterior surface of the tibia and to the fascia covering the muscle. Its tendon passes laterally superficial to the tendon of the tibialis posterior, and then runs behind the medial malleolus. The **flexor hallucis longus** is attached to the lower two-thirds of the posterior surface of the fibula, to the intermuscular septa and to the fascia over the muscle. Grooves for its tendon are found on the posterior surface of the talus and below the sustentaculum tali of the calcaneus. The **tibialis posterior** is attached to the interosseous membrane and to the adjacent areas of the tibia and fibula. Its tendon passes with the tendons of the flexor digitorum longus and the flexor hallucis longus behind the medial malleolus. The attachments of these muscles in the foot will be described later.

The three deep muscles plantar flex and invert the foot. The flexor hallucis longus and flexor digitorum longus also flex the toes. All the muscles of the back of the leg are supplied by the tibial nerve.

Examine the inferior tibiofibular joint, a fibrous joint in which the bones are united by a strong **interosseous ligament.** Posteriorly is the posterior tibiofibular ligament, the inferior transverse part of which is incorporated into the capsule of the ankle joint and is attached to the malleolar fossa of the fibula.

STRUCTURAL DETAILS

The tibia

The body of the tibia is triangular in cross-section except at its lower end where it is more circular. The upper end has been described with the knee joint. Attached to the tuberosity is the patellar ligament. The medial surface is smooth and subcutaneous, and no muscles arise from it. Traced downwards, this surface is continuous with the subcutaneous **medial malleolus** on the medial side of the ankle joint. The lateral side of the upper part of the bone is flattened by the attachment of the tibialis anterior. The posterior limit of the attachment of the muscle is marked by a sharp ridge, to which is attached the interosseous membrane, extending between the tibia and fibula.

Between the interosseous and medial borders is the posterior surface which is broad above where it joins the condyles. The soleal line, running downwards and medially across the upper third, marks the tibial attachment of the soleus, and below this line there is a large foramen for the nutrient artery. Above the soleal line is the area for the popliteus and below are the areas for the flexor digitorum longus medially and the tibialis posterior laterally. The lower end of the posterior surface is rounded and broadens above the ankle. Immediately posterior to the medial malleolus is a well-marked groove for the tendon of the tibialis posterior. On the lateral side of the lower end, the interosseous border widens to enclose a rough triangular area for the **interosseous ligament** (Fig. 48). This ligament is part of the fibrous inferior joint between the tibia and the fibula. The articular surface for the talus on the distal aspect of the body is slightly concave anteroposteriorly and extends medially over the whole of the lateral surface of the malleolus.

The fibula

This is a long and slender bone which has muscular attachments but does not transmit weight. Its upper end has a knob-like enlargement, the head, which articulates by a small flat facet with the posterolateral aspect of the lateral tibial condyle. Posterolateral to this facet of the **apex.** The body enlarges at its lower end to form the subcutaneous **lateral malleolus** which is long and pointed and extends about 2 cm beyond the tip of the medial malleolus. The medial surface of the lateral malleolus forms the lateral wall of the socket for the talus. Posterior to this articular surface is the **malleolar fossa.** Above the talar surface is the area for the interosseous ligament. On the posterior surface of the lateral malleolus is a well-marked groove for the tendons of the peroneal muscles (longus and brevis). The lateral surface of the malleolus is smooth and subcutaneous.

The body of the fibula has sharp borders which usually divide it into four surfaces ; these vary greatly in form with the muscular development of the individual. The arrangement of these surfaces is further complicated by a slight spiral twist of the body and need not be known in detail.

The calcaneus

The calcaneus is described here although the bone does not enter into the ankle joint but forms part of the talocalcaneal and calcaneocuboid joints (Fig. 55).

The posterior third of the calcaneus forms the prominence of the heel which is covered inferiorly by thick skin and fibrofatty tissue. The superior surface is distinguished by a rough posterior third markedly convex from side to side and three articular facets on the anterior two-thirds. The posterior facet is for the posterior calcaneal facet on the body of the talus, and the anterior pair are for the calcaneal facet on the head of the talus. On the calcaneus a shallow ligamentous groove between the posterior facet and the anterior facets is called the **sulcus calcanei.** This, together with the **sulcus tali,** forms the **sinus tarsi.** The middle of the three facets lies on the superior surface of a medial projection of the calcaneus, the **sustentaculum tali.** Below the sustentaculum is a groove for the flexor hallucis longus. The medial side of the body is rough and non-articular. The inferior surface is narrow and elongated anteroposteriorly. Its posterior end is enlarged to form a tuberosity. The lateral surface is flat and rough and has a small projection, the **peroneal trochlea,** at about its middle. The posterior surface is much larger than the anterior surface and the tendo calcaneus (tendo Achillis) is attached to its middle third. The anterior surface is relatively small and articulates with the cuboid.

Ossification (Fig. 44)

The epiphysis of the lower end of the tibia ossifies during the first year and joins the body at about the eighteenth year. The fibula has a centre for the lower end which appears in the second year and joins the body at about the eighteenth year. A primary centre for the calcaneus appears at about the seventh month of intra-uterine life.

The superior and inferior tibiofibular joints

There is little or no movement of the fibula on the tibia in man. The superior joint is synovial. The lower joint is fibrous with a strong **interosseous ligament**. In some fractures of the ankle, this ligament acts as a fulcrum and outward displacement of the fibular malleolus causes a fracture of the fibula above the ligament, which remains more or less intact.

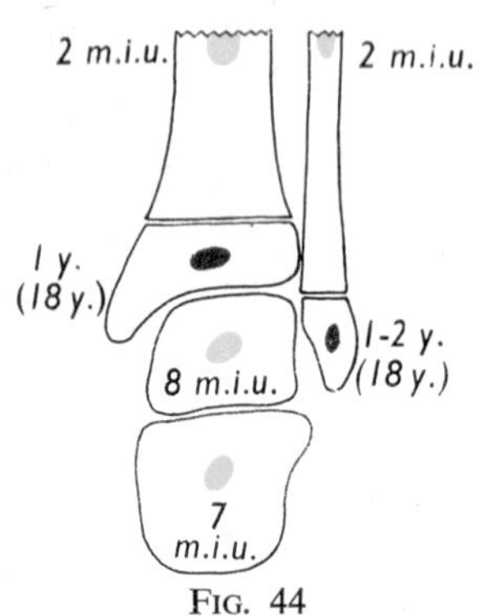

FIG. 44

The primary centres of ossification of the bones forming the ankle joint are shown in yellow and the secondary centres in red. The figures in brackets indicate the times of fusion of the primary and secondary centres (y., years; m.i.u., months in utero).

Muscles

The attachments of the two heads of the **gastrocnemius** and of the **plantaris** to the femur have already been described (page 133). The **soleus** has an extensive tendinous attachment to the upper part of the posterior surface of the fibula. It is also attached to the soleal line on the posterior surface of the tibia and forms a fibrous arch over the posterior tibial vessels. The adjacent surfaces of the gastrocnemius and the soleus are tendinous. The long narrow tendon of the plantaris fuses with the medial side of the common tendon, the tendo calcaneus, which is attached to the middle third of the posterior surface of the calcaneus. The gastrocnemius, soleus and plantaris are supplied by branches of the tibial nerve. The soleus, the deepest muscle, crosses the ankle joint only and is used in balancing the body on the foot. The gastrocnemius crosses the knee and ankle joints and can act on both. These muscles are very active in walking, going up stairs and down stairs, and climbing.

Vessels and nerves

The **popliteal artery** divides into its anterior and posterior tibial branches at the lower border of the popliteus. The **posterior tibial artery** passes straight down the calf and ends below and behind the medial malleolus by dividing into the **medial** and **lateral plantar arteries.** It lies deep to the soleus, on the tibialis posterior and between the flexor hallucis longus laterally and flexor digitorum longus medially, to all of which it gives branches. Its main branch, the **peroneal artery,** passes downwards and laterally between the fibula and flexor hallucis longus and ends in the anastomosis round the ankle joint and heel.

The **anterior tibial artery,** the other terminal branch of the popliteal artery, passes forwards through the upper part of the interosseous membrane into the anterior compartment of the leg. It turns downwards on the front of the membrane between the tibialis anterior medially and the other extensor muscles laterally. At the front of the ankle joint it is crossed anteriorly from its lateral to its medial side by the tendon of the extensor hallucis longus. It enters the foot where it becomes the **dorsalis pedis artery.** After running forwards over the dorsum of the tarsus (it can be felt pulsating in the living foot) it passes into the space between the bases of the first and second metatarsals, and from there into the sole of the foot, where it joins the lateral plantar artery and completes the plantar arch.

The **tibial nerve,** after giving branches to the gastrocnemius, soleus, plantaris and popliteus, passes deep to the upper edge of the soleus in the space between the two bones. It runs with the vessels as far as the medial malleolus. On its course down the leg it supplies the deep muscles of the calf, *i.e.* the tibialis posterior, flexor digitorum longus and flexor hallucis longus.

The **common peroneal nerve** winds round the neck of the fibula and enters the peroneus longus (Fig. 36) where it divides into two branches, the deep peroneal and superficial peroneal nerves. The **deep peroneal nerve** passes forwards through the fibres of the extensor digitorum longus and lies lateral to the anterior tibial artery. It supplies the two long extensors, the tibialis anterior and the peroneus tertius. Its terminal branch enters the foot in front of the ankle joint, supplies the extensor digitorum brevis and ends by supplying the skin of the first interdigital cleft. The **superficial peroneal nerve** runs downwards in the peroneal compartment deep to the peroneus longus. It supplies the peroneus longus and brevis, and emerges from the latter halfway down the leg to become cutaneous, supplying the lateral side of the leg, the contiguous sides of the 2nd, 3rd, 4th and 5th toes and the medial side of the great toe. The **sural nerve** is a branch of the tibial nerve, receives communicating branches from the common peroneal nerve and supplies the skin on the back of the calf, and on the lateral side of the ankle, foot and little toe. The **saphenous nerve** enters the foot in front of the medial malleolus and can be followed forwards on the medial side of the foot as far as the metatarso-phalangeal joint.

FUNCTIONAL ASPECTS

The **soleal pump** is the name given to the system of large veins found in and between the muscles of the calf. The muscles and vessels are all enclosed in a relatively non-elastic fibro-osseous space. Contraction of the muscles forces the blood in the veins up into the popliteal vein and thus assists venous return. If the valves, which prevent blood passing from the deep to the superficial veins, are incompetent the blood may be forced into the superficial veins which may become varicose.

CHAPTER 20

THE ANKLE JOINT

INTRODUCTION

THE ankle is a relatively simple hinge joint which has to carry the weight of the body. Of all the joints in the lower limb, the ankle is most liable to injury and to dislocation.

The bones involved are the tibia, fibula and talus. The former two are shaped and held together as a mortise which embraces the trochlea of the talus. The talus articulates with the calcaneus inferiorly and the navicular distally.

DISSECTION

Before proceeding to the dissection of the ankle joint, re-examine the deep fascia in this region. It is thickened on three aspects of the joint to form **retinacula** (Fig. 45).

1. The **extensor retinacula.** The upper band passes between the tibia and fibula above the ankle joint. The lower is shaped like the letter Y with the stem attached laterally to the calcaneus and the two limbs to the medial side of the foot.

2. The **flexor retinaculum.** This band passes between the medial malleolus and the medial side of the calcaneus.

3. The **peroneal retinacula.** The upper band passes between the lateral malleolus and the calcaneus laterally. The lower band is attached to the calcaneus above and below the peroneal trochlea.

These retinacula retain the tendons in position, especially in dorsiflexion of the foot. With the bones, they form canals in which the tendons run. In this part of their course the tendons are surrounded by synovial sheaths.

Pull away all the tendons round the ankle joint after incising the retinacula, and turn them distally in order to get good access to the region. Plantar flex the foot and expose the **capsule** anteriorly. It is thin and is attached to the lower end of the tibia and to the neck of the talus. On the medial side of the joint expose the triangular **medial (deltoid) ligament,** a strong thickened part of the capsule (Fig. 47). Examine its attachment to the tibia above, which is the apex of the triangle. Below, the base is attached to the neck of the talus, the navicular, the ligament between the navicular and sustentaculum tali, the sustentaculum tali, and again to the talus.

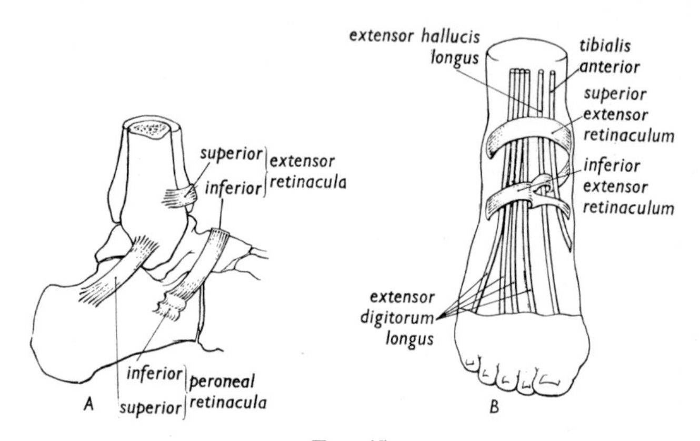

FIG. 45
The attachments of the extensor and peroneal retinacula are indicated.

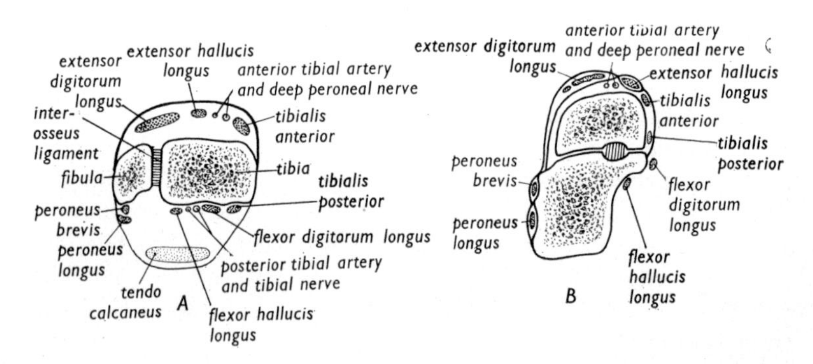

FIG. 46
A shows the principal structures seen in a horizontal section above the ankle joint and B is an oblique section through the talus and calcaneus below the ankle joint.

On the lateral side three thickenings in the capsule radiate from the lower end of the lateral malleolus. Dissect them out and examine their attachments. They are from front to back, (1) the

anterior talofibular ligament, running to the neck of the talus, (2) the **calcaneofibular ligament** running downwards and backwards to the side of the calcaneus, and (3) the **posterior talofibular ligament,** running to the lateral tubercle of the talus. Note that the third ligament is attached to the malleolar fossa and is entirely posterior to the ankle joint (Fig. 48B).

Dorsiflex the foot and examine the posterior part of the joint. Here there is a very thin, lax capsule to which are attached the lower fibres of the transverse part of the posterior tibiofibular ligament. Also strengthening this aspect of the joint are the fibres of the posterior talofibular ligament.

Finally, cut through the anterior part of the capsule by a transverse incision. Examine the articular surfaces of the bones, gradually extending the incision backwards by cutting through the medial and lateral ligaments. When the joint is fully exposed dorsiflex and plantar flex the foot and note the following points. (1) The talus fits closely into the socket formed by the tibia and the malleoli. The shape of this socket permits dorsiflexion and plantar flexion. (2) The front of the upper articular surface of the talus is broader than the back ; therefore in dorsiflexion it is gripped more firmly by its socket. (3) The posterior part of the articular surface of the tibia is continued on to the deep surface of the transverse part of the posterior tibiofibular ligament. This ligament, then, assists in forming the socket. Completely detach the foot from the leg by cutting through the remaining ligaments and other structures round the joint.

STRUCTURAL DETAILS

The lower ends of the tibia and the fibula have already been described on page 156. The superior articular surface of the ankle joint is formed by the deep surfaces of the malleoli and the distal surface of the tibia (Fig. 48).

The talus

The talus fits into the mortise formed by the tibia and lateral malleolus of the fibula, and has large articular surfaces, small ligamentous areas and no muscular attachments. The superior surface for the lower end of the tibia is easily identified. It is

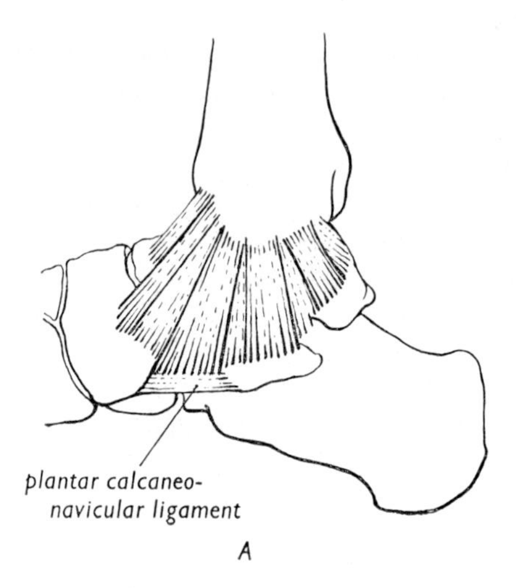

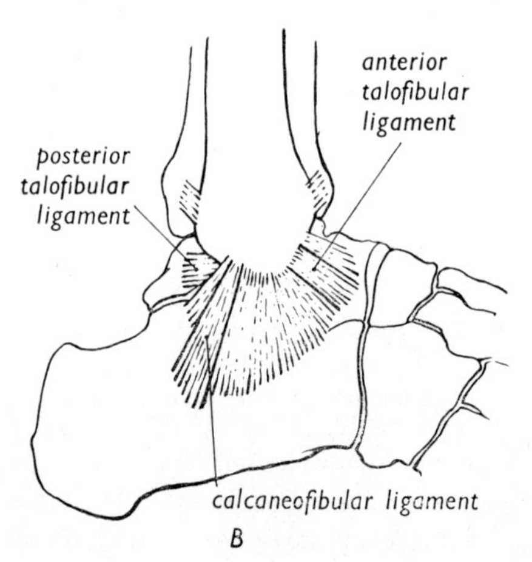

FIG. 47

Drawings of the principal attachments of the medial ligament (A) and of the lateral ligament (B) of the ankle joint. The parts of the lateral ligament are named according to their bony attachments.

markedly convex anteroposteriorly, and very slightly concave from side to side. This surface is broader anteriorly than posteriorly and is continuous both laterally and medially with the articular surfaces

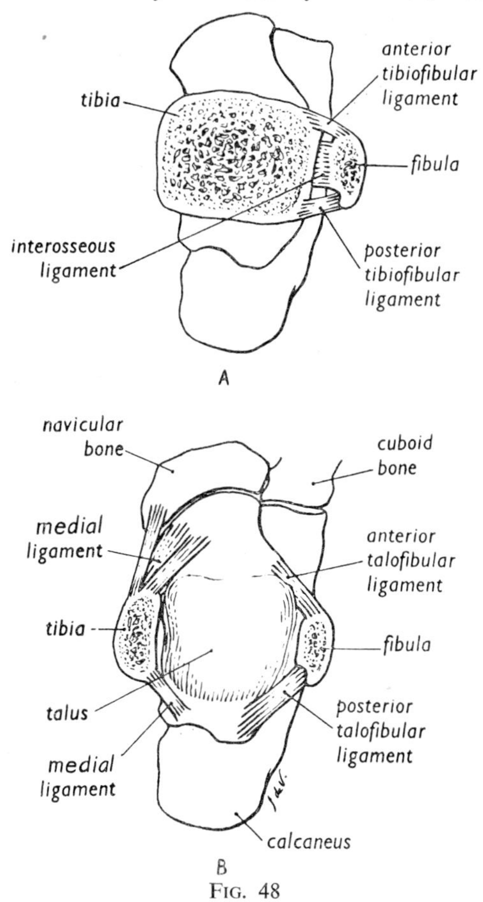

FIG. 48
In A, the lower ends of the tibia and fibula have been sectioned just above the ankle joint. In B the section passes through the joint, the tips of the malleoli being cut.

for the malleoli. The lateral surface is much longer than the medial, corresponding to the larger articular area on the medial surface of the fibular malleolus, and is triangular with the apex downwards. The medial surface, for the tibial malleolus, is comma-shaped. The three articular surfaces are called the **trochlea** of the talus.

In front of the trochlea the bone is rough and somewhat constricted, forming the **neck,** and more distally is the **head.** The neck, if followed laterally, leads to the deep **sulcus tali** on the inferior surface. Posterior to the sulcus tali there is an elongated concave facet for the calcaneus. Posteromedial to this facet is a groove for the tendon of the flexor hallucis longus. This groove divides the posterior process into medial and lateral tubercles.

Anterior to the sulcus tali is the rounded articular surface of the head (convex from side to side and from above downwards). This large surface articulates with the calcaneus proximally, the plantar calcaneonavicular (spring) ligament in the middle, and the navicular distally (Fig. 55).

The structure of the ankle joint

The ankle joint is a synovial hinge joint between the articular areas at the lower ends of the tibia and fibula and the trochlea of the talus. The tibia and fibula, bound firmly together by the interosseous ligament, form a socket for the talus. They grip it tightly, especially in dorsiflexion, and the shape of the socket prevents any active movements other than plantar flexion and dorsiflexion since the anterior part of the trochlea is broader than the posterior part. The shape of the articular surfaces thus contributes considerably to the stability of the joint which has to carry the whole of the superimposed body weight.

The **capsule** has thin anterior and posterior ligaments. The lateral and medial portions of the capsule are reinforced by powerful ligamentous thickenings as would be expected with the movements found in the ankle joint. The anterior ligament is lax ; it is attached above to the anterior margin of the articular surface of the tibia, and below to the neck of the talus. The posterior ligament, also thin and lax, is attached above to the articular margin of the tibia and below to the posterior edge of the articular surface of the talus.

The **medial (deltoid) ligament** is a strong band continuous with the capsule (page 161), and prevents excessive bending of the foot outwards on the leg. The **lateral ligaments** consist of three separate bands (page 162), and prevent excessive inward bending of the foot on the leg.

The **synovial membrane** lines the internal surface of the capsule and the ligaments, and is attached to the bone close to the articular margins.

The **relations** of the joint are, anteriorly, the dorsiflexor tendons, the anterior tibial vessels and the deep peroneal nerve. Posteriorly a pad of fat separates the joint from the tendo calcaneus. Postero-medially are the tendons of the long flexor muscles, the posterior tibial vessels and the tibial nerve. Posterolaterally are the tendons of the evertor muscles, peroneus longus and brevis (Fig. 46).

FUNCTIONAL ASPECTS

The movements at the ankle joint

In the active movements of dorsiflexion and plantar flexion, the talus rotates round a transverse axis passing through the tip of the lateral malleolus and just below the tip of the medial malleolus. The anatomical position is defined as that in which the foot is at right angles to the long axis of the tibia and the toes are directed forwards. Plantar flexion from this position is greater than dorsi-flexion, the total range being about 40°. There is considerable individual variation.

Plantar flexion is produced by the soleus and gastrocnemius, aided by the long flexors of the toes, the tibialis posterior and the peroneus longus and brevis. It is checked by the dorsiflexors of the foot and the anterior part of the capsule of the ankle joint.

Dorsiflexion is produced by the tibialis anterior, the long extensors of the toes and the peroneus tertius. Because the line of weight passes in front of the ankle joints in the anatomical position, the body tends to fall forwards at the ankle joints and this is con-trolled by the calf muscles. The two sets of muscles, at the front and the back of the leg, serve to balance the body on the foot. The calf muscles during walking are said to raise the heel so that the whole body weight is thrown forwards (page 190). They are especially important in walking uphill or upstairs.

CHAPTER 21

THE FOOT

INTRODUCTION

THE foot of man is adapted to allow the weight of the body to be distributed to the ground and also to be used as a lever with which the body is propelled forward. For the function of supporting the weight, the foot is arched longitudinally and transversely. The medial side of the longitudinal arch consists of the talus at the apex, the calcaneus behind, and the navicular, three cuneiforms and three medial metatarsals in front (Figs. 49 and 57). The lateral side of the arch is lower and consists of the calcaneus, cuboid and two lateral metatarsals (Figs. 49 and 58). The weight is distributed to the heel and to the metatarsal heads. The transverse arch has the effect of converting the foot into a dome-like structure with the talus at the apex. The muscles and ligaments of the foot are arranged so as to maintain these arches and allow them some degree of adjustment. The foot can also be inverted and everted about a longitudinal axis (cf. the movements of the palm of the hand in supination and pronation). These movements of the foot take place chiefly between the talus and the rest of the tarsus.

Both the intrinsic and extrinsic muscles of the foot are important in helping the ligaments to maintain the arches and in adjusting the foot to allow it to work as a lever and to carry the body over uneven ground. To work as a lever the foot has a certain rigidity, but it is also resilient enough not to break under the strain. In walking, the force is applied at the calcaneus by the tendo calcaneus, the weight is carried on the talus and the fulcrum is at the heads of the metatarsals. The only joints at which bending must occur during walking on a flat surface are the ankle and metatarsophalangeal joints. In walking the toes are pressed on to the ground to increase friction as the foot lever is raised. The muscles of the metatarsophalangeal joints, particularly those acting on the hallux, take a considerable part of the strain at one stage of walking. For this purpose the hallux is held parallel to the other digits and is not opposable as it is in arboreal primates.

Identify again the separate bones of the foot on an articulated skeleton and note how the arches are formed and that the first metatarsal bone is much stronger than any of the others (Fig. 49).

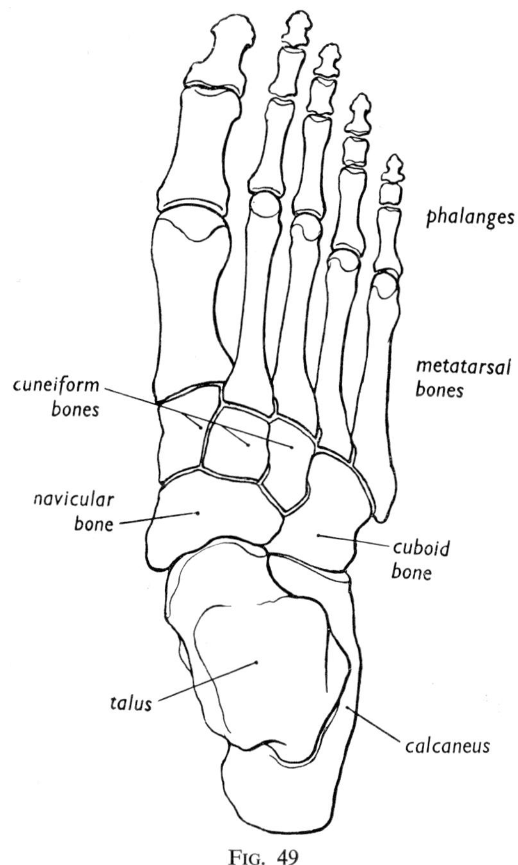

FIG. 49
The bones of the foot.

DISSECTION

Although instructions for dissection follow, it is suggested that it may not be necessary to dissect the sole of the foot. Remove the skin. Over the sole it is firmly bound to the underlying fascia (the plantar aponeurosis) by numerous fibrous septa which split up the fat into many lobules, much smaller than elsewhere in the body. The skin is thickened beneath the weight-bearing areas of the heel and the heads of the metatarsals. Remove the fat and superficial fascia and clean the surface of the underlying **plantar aponeurosis.** Examine its posterior attachment to the tuberosity of the calcaneus. Define its three portions, the thick central portion running to the five toes and dividing distally to ensheath the flexor tendons of the toes, and the thinner medial and lateral portions covering the small muscles of the 1st and 5th toes respectively.

Cut away the posterior attachment of the aponeurosis from the calcaneus and turn it forwards, separating it from the deeper structures. It will be necessary to cut it away from two septa passing from its deep surface into the deeper parts of the foot. These intermuscular septa separate a large central space from smaller medial and lateral spaces.

The first layer of muscles (Fig. 50)

Identify and clean the following muscles lying deep to the plantar aponeurosis: 1. flexor digitorum brevis in the centre, 2. abductor hallucis, medially, 3. abductor digiti minimi, laterally.

Posteriorly these muscles are all attached to the tuberosity of the calcaneus and to the deep surface of the plantar aponeurosis. The **abductor hallucis** is attached anteriorly to the medial side of the proximal phalanx of the great toe with the medial tendon of the flexor hallucis brevis. The **flexor digitorum brevis** divides into four tendons each of which splits to allow the flexor digitorum longus to pass through, and the two slips are attached to the sides of the middle phalanx of the 2nd to the 5th toes. The **abductor digiti minimi** is attached to the lateral side of the base of the proximal phalanx of the 5th toe.

These muscles assist in maintaining the longitudinal arches of the foot. The actions which their names suggest are relative to a line along the length of the 2nd toe.

Cut through the muscles close to their origins from the calcaneus and turn them forwards. The structures between the first and the second layer of muscles are now exposed (Fig. 51). Pick up the lower ends of the posterior tibial artery and the tibial

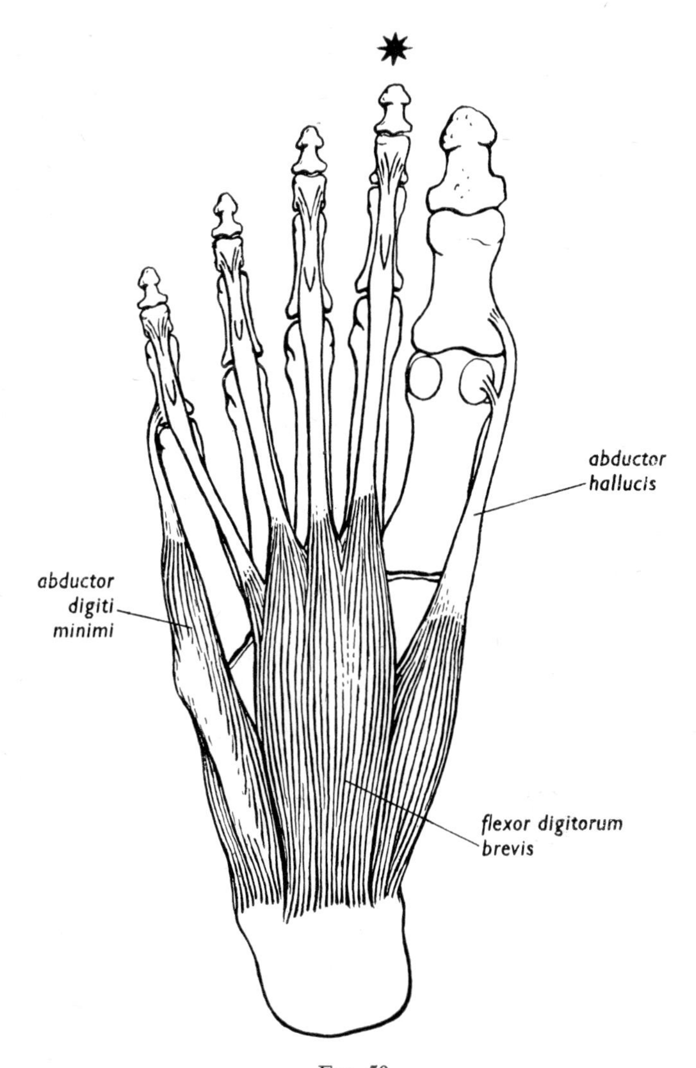

FIG. 50
The first muscular layer of the foot. (Abduction and adduction of the toes are related to an axis line through the second toe marked *.)

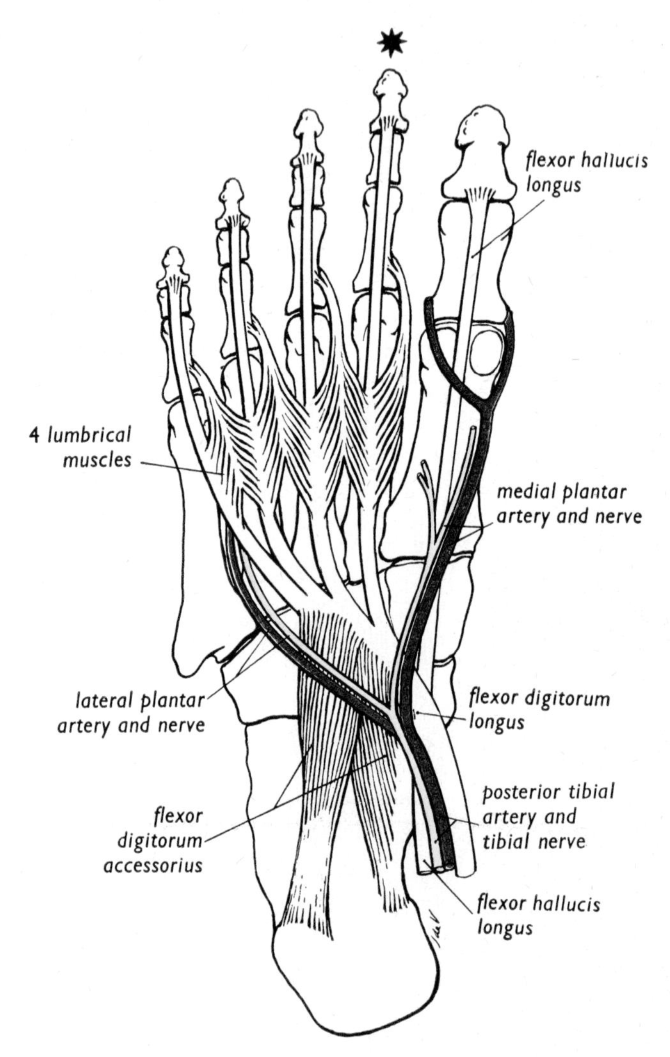

FIG. 51

The second muscular layer of the foot and the division of the posterior tibial artery and the tibial nerve into the medial and the lateral plantar arteries and nerves.

nerve which were seen behind the medial malleolus. Trace them forwards below the sustenaculum tali. They divide into lateral and medial plantar branches (Fig. 51). The **medial plantar nerve** runs along the medial side of the foot and supplies four digital cutaneous branches for the medial three and a half toes, and muscular branches to the abductor hallucis, the flexor hallucis brevis, the flexor digitorum brevis and the medial lumbrical. The **lateral plantar nerve** runs obliquely across the foot towards the little toe. It supplies the lateral one and a half toes with cutaneous branches and passes among the deeper structures to supply all the remaining muscles in the sole of the foot. Compare the distribution of the medial plantar nerve with that of the median nerve, and the lateral plantar nerve with that of the ulnar nerve.

The posterior tibial artery ends below the medial malleolus by dividing into lateral and medial plantar arteries. The **medial plantar artery** accompanies the medial plantar nerve and ends by supplying the medial side of the great toe and anastomosing with the terminal branches of the lateral plantar artery.

The **lateral plantar artery** runs laterally with the lateral plantar nerve between the first and second layers of muscles to the base of the fifth metatarsal. Here it turns medially and deeply and will be dissected later.

The second layer of muscles (Fig. 51)

Follow the tendon of the **flexor hallucis longus** from the back of the medial malleolus forwards to its insertion into the base of the terminal phalanx of the great toe. Trace the tendon of the **flexor digitorum longus** forwards from the region of the sustentaculum tali into the sole where it crosses superficial to the hallucis tendon and then splits into four tendons for the lateral toes. Find the lumbrical muscles arising from the flexor digitorum tendons. Follow the main flexor tendons with the tendons of the flexor digitorum brevis into their fibrous sheaths one of which should be split to show the actual insertion of the tendons into the phalanges. Trace the lumbrical tendons round the medial sides of the toes into the dorsal extensor expansion.

Examine the main tendon of the flexor digitorum longus before it divides into four and find the **flexor accessorius muscle.** This arises from the medial and the plantar surfaces of the calcaneus; it is inserted into the lateral side of the flexor digitorum longus tendon.

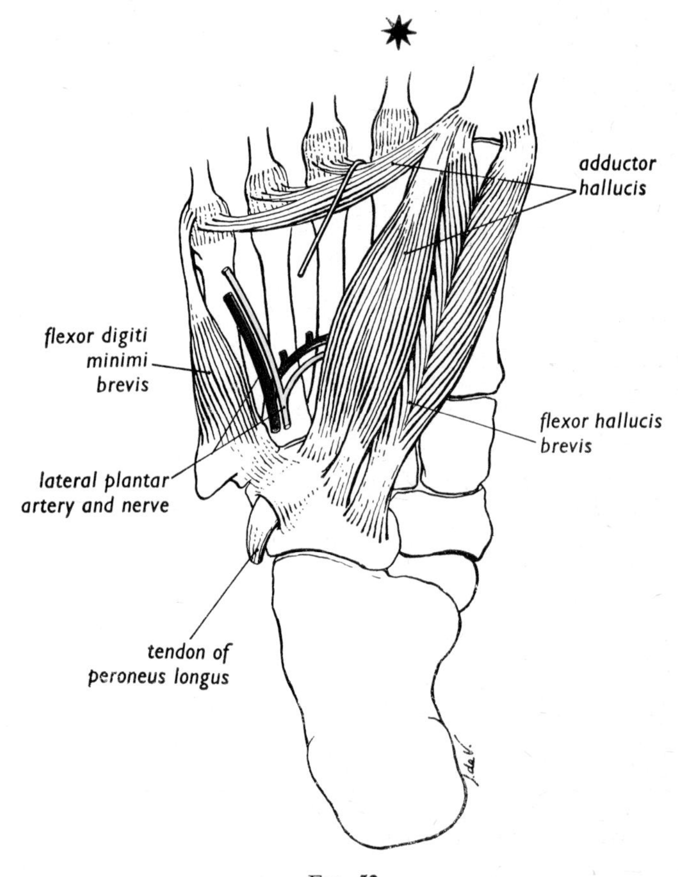

FIG. 52
The third muscular layer of the foot.

The lumbrical muscles help to flex the metatarsophalangeal joints and extend the interphalangeal joints. The accessorius converts the oblique pull of the flexor digitorum longus into a straight backward pull and can also maintain the action of the longus muscle when the muscle fibres in the calf are relaxed.

Cut through the long flexor tendons and the calcaneal attachments of the flexor accessorius and expose the deeper layers (Fig. 52).

The third layer of muscles (Fig. 52)

Identify three muscles. On the lateral side, the **flexor digiti minimi brevis** is attached distally to the base of the proximal phalanx, and more medially, the **flexor hallucis brevis** divides into two parts passing to the medial and lateral sides of the base of the proximal phalanx. Close to its distal attachment each tendon contains a sesamoid bone which articulates with the plantar surface of the 1st metatarsal head. In the groove between the two sesamoids lies the long flexor hallucis tendon. Both the short flexor muscles arise from the region of the plantar surface of the cuboid. Expose the **adductor hallucis muscle.** This has an oblique head attached close to the two short flexors, and a transverse head attached to the ligaments of the metatarsophalangeal joints. Both heads are attached, with the lateral head of the flexor hallucis brevis, into the base of the proximal phalanx of the great toe.

The transverse head of the adductor helps to prevent spreading of the heads of the metatarsals when they bear weight. The other muscles help support the longitudinal arches of the foot. Cut away the short muscles from their proximal attachments and turn them forwards, so exposing the deepest layer.

The fourth layer of muscle (Fig. 53)

Examine the tendons of the peroneus longus and tibialis posterior. The **peroneus longus** is seen turning round the lateral border of the foot where it enters a groove on the plantar surface of the cuboid. This groove is converted into a tunnel by the **long plantar ligament,** which is a strong, broad band passing from the plantar surface of the calcaneus to the edges of the groove on the cuboid and then on to the bases of the middle three metatarsals. Find the peroneus longus tendon emerging from the medial end of the tunnel and trace it to its attachment to the lateral side of the base of the 1st metatarsal and medial cuneiform. Passing under the foot in this way the tendon supports the lateral side of the arch and everts the foot.

Grasp the tendon of the **tibialis posterior** at the level of the medial malleolus and dissect it below the plantar calcaneonavicular ligament to its distal extensive attachments in the foot. Try to

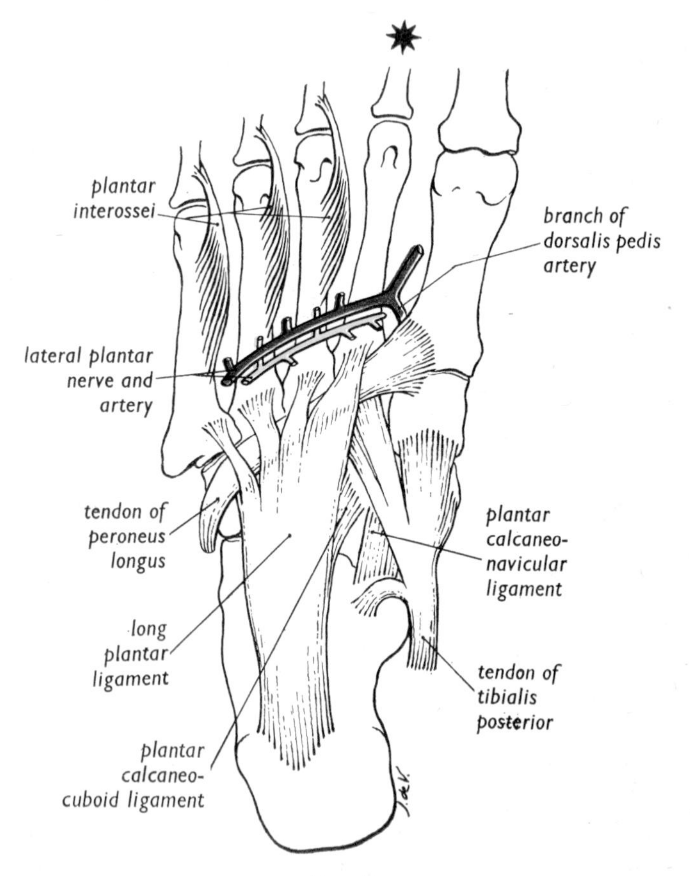

FIG. 53
The fourth muscular layer of the foot and the lateral plantar artery and nerve.

demonstrate three important bands, (1) an attachment to the tuberosity of the navicular on its plantar surface, (2) a recurrent slip passing backwards and laterally to the anterior surface of the sustentaculum tali, (3) a band passing forwards to the cuboid and cuneiform bones. The muscle supports the medial side of the longitudinal arch and inverts the foot.

Follow the lateral plantar artery as it turns medially across the sole. It forms the plantar arch and, at the proximal end of the first intermetatarsal space, it anastomoses with the dorsalis pedis artery. The deep branch of the lateral plantar nerve accompanies the artery across the foot and supplies most of the deeply placed muscles. Digital branches of the artery and nerve supply the toes.

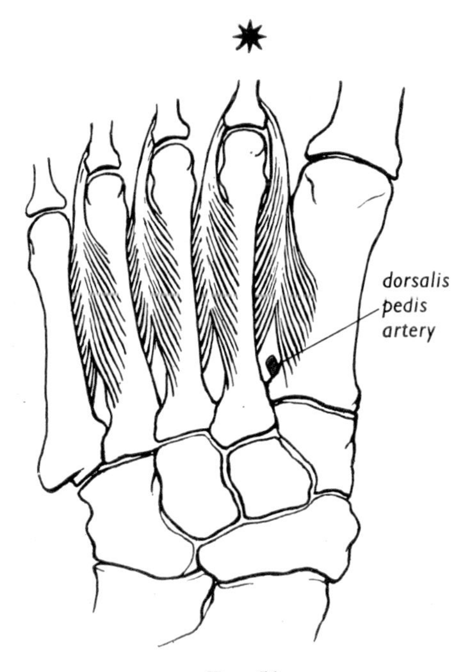

FIG. 54
The dorsal interossei.

Examine the spaces between the metatarsal bones and display the three **plantar interosseus muscles.** These muscles are attached to the shafts of the three lateral metatarsals and pass to the medial side of the base of the corresponding proximal phalanx. To see them properly, cut through the deep transverse ligaments binding the heads of the metatarsals together. Remove the plantar interossei in order to examine the four **dorsal interossei** (Fig. 54). One is attached distally to the medial and one to the lateral side of the 2nd toe, and the other two are attached to the lateral side of the 3rd and 4th toes. Unlike the plantar interossei, they arise from both bones bounding an intermetatarsal space. These muscles reduce the lateral spreading, which tends to occur during weight bearing.

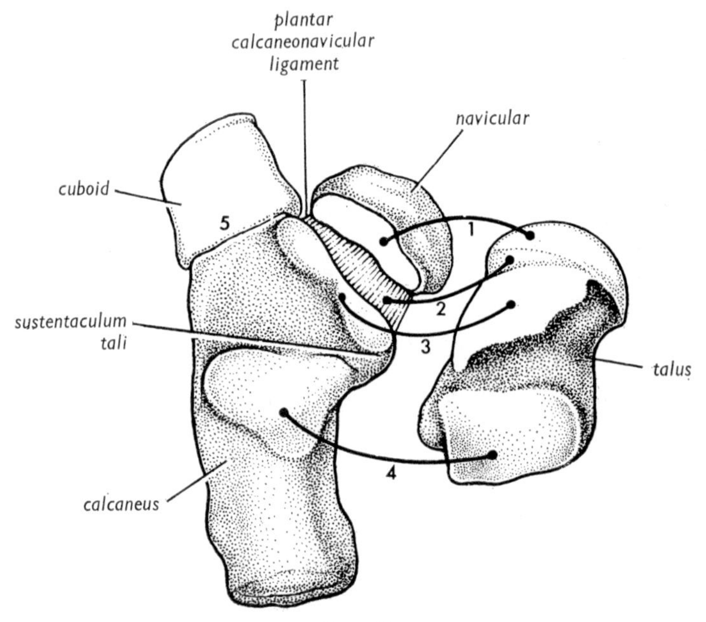

FIG. 55
Diagram showing the articulating surfaces at the talocalcaneonavicular joint (1, 2 and 3), the subtalar joint (4) and the calcaneocuboid joint (5).

The joints of the foot

Having completed the dissection of the sole of the foot strip off all the soft tissues except the distal portions of the long tendons. The talus is still in position on top of the calcaneus. Examine its relation to the bony prominences of the calcaneus and note its obliquity and the way in which its anterior portion seems to overhang the medial side of the foot.

Cut through the dorsal part of the talonavicular capsule, thus opening up the talonavicular joint. Insert a narrow blade into the sinus tarsi, the tunnel between the talus and the calcaneus, and divide the interosseous talocalcaneal ligament which holds the two bones together. Remove the talus from the top of the calcaneus by cutting any capsule remaining between them.

Examine the articular surfaces exposed (Fig. 55). Identify the posterior facet on the calcaneus for the body of the talus (4), and the middle and the anterior facets for the head of the talus (3). Examine the concave surface on the proximal aspect of the navicular for the head of the talus (1). Between the plantar aspect of the navicular and the sustentaculum tali stretches the **plantar calcaneonavicular (spring) ligament.** Replace the talus in position and viewing it from the medial side, note how the head of the talus is directly in contact with the upper surface of the spring ligament (2). Also note how the middle and anterior calcaneal facets, the upper surface of the spring ligament and the hollow of the navicular, form one continuous socket for the reception of the head of the talus (Fig. 55).

On the plantar surface of the foot examine the spring ligament and note its attachments to the sustentaculum tali and the navicular. Also note that the tendon of the tibialis posterior as a whole lies directly below the ligament; the significance of this will be appreciated in the study of the arches of the foot. Further laterally, identify and examine the **long plantar** and **plantar calcaneocuboid ligaments** (Fig. 53). The former runs from the plantar surface of the calcaneus to the edge of the groove on the plantar surface of the cuboid and then on to the bases of the 2nd, 3rd and 4th metatarsal bones. The latter runs obliquely from the plantar surface of the calcaneus to the ridge on the cuboid. If it is not readily seen, cut through the long plantar ligament just proximal to the bases of the metatarsals and pull it backwards. This will expose the plantar calcaneocuboid ligament and the tendon of the peroneus longus, which should be traced to its attachments on the base of the first metatarsal and the medial cuneiform bones.

The foregoing are the only ligaments which should be dissected and recognised. Note that in the foot there are also strong interosseous ligaments between many of the smaller bones, and that the plantar ligaments are much thicker and stronger than the dorsal.

The articulation between the talus and navicular lies in the same transverse plane as the articulation between the calcaneus and cuboid. Although there are two joints with separate synovial cavities, they are known as the **transverse tarsal** (midtarsal) **joint.** A small amount of inversion and eversion of the foot can take place at this joint.

STRUCTURAL DETAILS

The **navicular bone** has a large concave facet on its proximal surface for the head of the talus. On the medial side of the bone is a prominent tuberosity to which part of the tibialis posterior is attached. The anterior surface has three articular facets for the cuneiform bones. The **medial cuneiform** is the largest of the three wedge-shaped cuneiforms, its superior (dorsal) surface is narrow and its inferior (plantar) surface broad. The intermediate cuneiform is the smallest of the three, and has its base superiorly, as has the lateral cuneiform. In the articulated foot, the proximal surfaces of the three cuneiforms form a slightly concave surface for the navicular. They articulate with each other and the lateral cuneiform articulates with the cuboid (Fig. 49). Articulating with the anterior surface of the calcaneus is the **cuboid.** It has a deep groove for the tendon of the peroneus longus on its inferior surface. It articulates with the calcaneus proximally, the lateral cuneiform (and occasionally with the navicular) medially, and the 4th and 5th metatarsals distally. The **metatarsals** are longer than the metacarpals and their shafts are compressed from side to side. Their inferior surface is concave. The bases articulate with the distal tarsal bones. The heads are convex. Their articular surfaces extend on to the dorsal and plantar aspects and articulate with the proximal phalanges. The 1st metatarsal is shorter and much thicker than the rest and has usually no articulation with the 2nd metatarsal. The 2nd metatarsal articulates at its base with the medial, intermediate and lateral cuneiforms as well as with the 3rd metatarsal. The base of the 5th metatarsal has an elongated **tuberosity** projecting proximally on its lateral side.

There are two **phalanges** in the 1st toe and three in each of the other toes. The terminal phalanges are small and their distal ends are broadened for the nail beds.

Ossification (Fig. 56)

The tarsal bones are each ossified from one primary centre. Three of these centres are present at birth—the calcaneal, talar

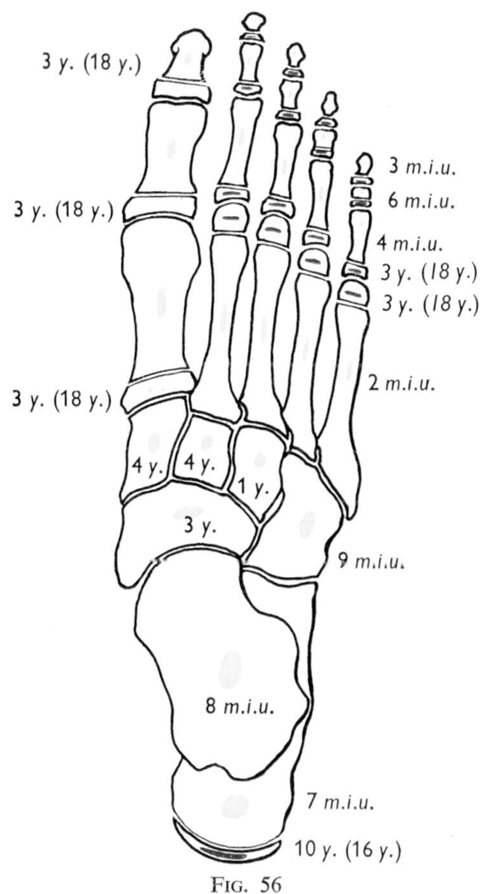

FIG. 56
The primary centres of ossification of the bones of the foot are shown in yellow and the secondary centres in red. The figures in brackets indicate the times of fusion of the primary and secondary centres (y., years; m.i.u., months in utero).

and cuboid. The lateral, medial and intermediate cuneiforms and the navicular ossify from centres appearing between the first and fourth years. There is a scale-like epiphysis on the posterior surface of the calcaneus, and the posterior process of the talus sometimes has a separate centre of ossification.

The metatarsals and phalanges all have primary centres for the body, appearing at about the end of the second month of intra-uterine life. All the phalanges and the 1st metatarsal have an epiphysis at the proximal end which appears about the third year. The other metatarsals have an epiphysis at the distal end which appears about the same time. All these epiphyses fuse with the bodies at about the eighteenth year.

The joints of the foot

A detailed knowledge of the numerous joints in the foot is not necessary. The movements of dorsiflexion and plantar flexion at the ankle joint have been dealt with on page 167. Inversion is most easily performed with the foot in plantar flexion and eversion in dorsiflexion. The joints involved in these movements are the talocalcaneal, the talonavicular and the calcaneocuboid. Plantar flexion and dorsiflexion take place at the metatarsophalangeal and interphalangeal hinge joints.

The foot is designed for support as well as movement and for this purpose the bones are arranged in a system of arches. Most of the joints have flat articulating surfaces allowing slight gliding movements which give resilience to the arch. As the individual bones must be prevented from splaying out under the pressure of the body weight from above, there are strong plantar and interosseous ligaments.

The talocalcaneal joints (Fig. 55)

There are two of these, the **subtalar joint** behind the sinus tarsi, and part of the **talocalcaneonavicular joint** in front. The subtalar joint is between the concave facet on the lower surface of the talus and the convex posterior facet on the upper surface of the calcaneus. There is a weak capsule surrounding the joint but the main ligament uniting the bones is the interosseous talocalcaneal ligament in the sinus tarsi.

The talocalcaneonavicular joint

This is the joint between the ball formed by the head of the talus and the socket formed by the upper surface of the sustentaculum tali and the proximal surface of the navicular bone.

Filling the triangular gap between the navicular and the sustentaculum tali and attached to both is the **plantar calcaneonavicular (spring) ligament.** The head of the talus rests directly on the upper surface of this ligament which forms part of the joint surface. The joint cavity is surrounded by a capsule with a weak dorsal ligament passing from the talus to the navicular. The synovial membrane lines the non-articular surfaces.

The calcaneocuboid joint

This is between the saddle-shaped facets on the front of the calcaneus and the back of the cuboid. Dorsally, medially and laterally the ligaments connecting the two bones are weak. On the plantar surface are the strong long plantar and plantar calcaneocuboid ligaments.

The metatarsophalangeal joints

These are hinge joints with a strong plantar and a weak dorsal ligament. Binding the heads of the metatarsals together and attached to the plantar ligaments, are the deep transverse ligaments, which prevent splaying of the toes when weight is carried by the foot. The metatarsophalangeal joint of the great toe is slightly different from the others. The plantar ligament is relatively thicker and stronger and embedded in it are the two sesamoid bones in the tendons of the flexor hallucis brevis, which articulate with the plantar surface of the head of the first metatarsal. These bones bear weight and thus protect the tendon of the flexor hallucis longus which lies in the hollow between the bones. The movements that take place at these joints are plantar flexion and dorsiflexion and possibly some abduction and adduction.

FUNCTIONAL ASPECTS

The movements of the foot

Inversion and eversion take place at the subtalar joint and the talocalcaneonavicular joint. The talus remains fixed in both movements. In inversion, the sole of the foot is turned inwards and the medial border is moved medially and upwards. In eversion, a much more limited movement, the sole is turned outwards and the lateral border moves upwards and outwards.

The cuboid, the cuneiforms and the forepart of the foot move with the navicular and calcaneus. There is also a slight amount of gliding at the calcaneocuboid joint.

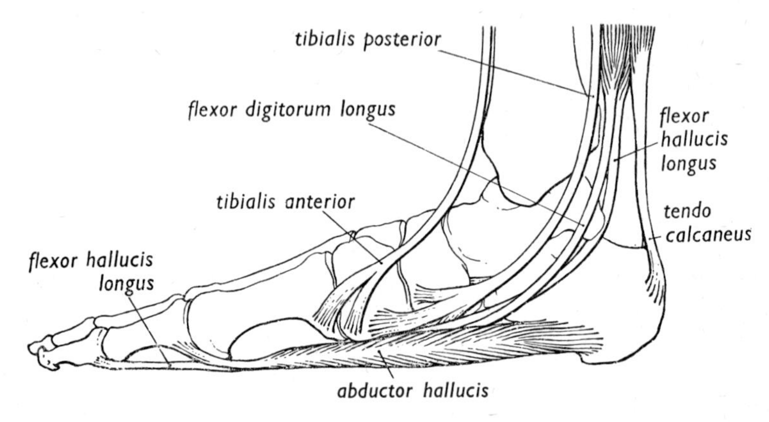

FIG. 57
Diagram of the medial side of the longitudinal arch.

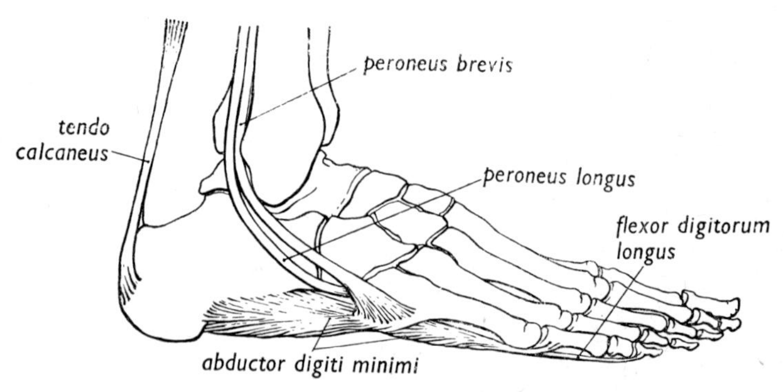

FIG. 58
Diagram of the lateral side of the longitudinal arch.

The muscles producing inversion are the tibialis anterior and tibialis posterior, assisted to a slight extent by the long flexors of the toes (Fig. 57). Eversion is produced by the three peroneal muscles (Fig. 58). The movements are checked by the interosseous ligament between the talus and calcaneus and by the medial and lateral ligaments of the ankle. Inversion is maximal in plantar flexion and eversion in dorsiflexion.

The movements of the toes

Flexion and extension take place at the metatarsophalangeal and interphalangeal joints. The muscles producing these movements are the long and short flexors and extensors. Abduction and adduction are possible at the metatarsophalangeal joints. These movements are not important in people who wear shoes.

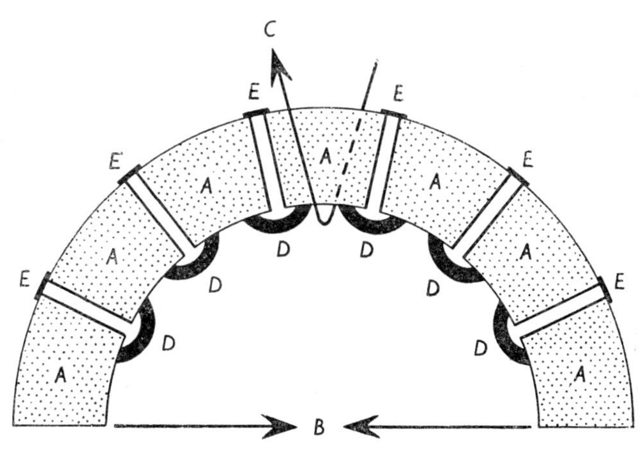

FIG. 59

Diagram to show how an arch, such as is found in the foot, may be maintained. Note (A) the shape of the component segments, (B) the ties holding the ends of the arch together, (C) the long slings supporting the top of the arch, (D) the strong intersegmental ties on the concave surface and (E) the weaker ties in the convex surface.

The arches of the foot (Figs. 57 and 58)

Although the foot is used as a lever in propelling the body forward, it is not a rigid structure. It receives and absorbs the impact of the body weight on its upper surface when the foot comes into contact with the ground. Resilience is imparted to the foot by its construction in the form of arches which are maintained by ligamentous and muscular ties and braces. There are longitudinal and transverse arches in the foot.

In the support of the arches of the foot three factors play a part: (1) the shape of the bones, (2) the ties of the ligaments and short muscles, and (3) the bracing or slinging action of the long muscles. If we consider an arch of wedge-shaped blocks, as in Figure 59, the shape of the blocks (A) will make the arch stable if the ends of the arch are kept in position. This can be done by a tie (B) between the ends of the arch. The centre of the arch can be further maintained by a brace (C) passing under it from above and acting as a sling. The separate elements of the arch can also be tied together on their concave side by strong individual ties (D).

The longitudinal arch

The posterior base of the arch is formed by the tuberosity of the calcaneus and the anterior base by the heads of the metatarsals. The metatarsal and tarsal bones form the body of the arch, the talus being its highest member. The lateral part of the arch, composed of the calcaneus, cuboid and lateral two metatarsals, is flatter than the medial part, which is composed of the calcaneus, talus, navicular, cuneiforms and medial three metatarsals (Figs 57 and 58).

The arch is maintained by ligaments and muscles. The ligamentous supports are: (*a*) the plantar calcaneonavicular ligament, which supports the head of the talus and the medial side of the arch, (*b*) the plantar calcaneocuboid ligament tying together the calcaneus and cuboid and therefore supporting the lateral part of the arch, (*c*) the long plantar ligament, tying together the calcaneal tuberosity and the bases of the lateral metatarsals, and supporting the lateral part of the arch, (*d*) the intertarsal ligaments on the plantar side, (*e*) the plantar aponeurosis, tying the two ends of the arch.

The muscular supports of the arch are: (*a*) the tibialis posterior, which is attached to the navicular and to the bones of the anterior part of the arch. (A special recurrent slip passes backwards to the sustentaculum tali, lying directly underneath the spring ligament and therefore beneath the head of the talus, an example of the interrelationship of muscular and ligamentous support), (*b*) the tibialis anterior, which supports the front of the arch by its attachment to the medial surface of the medial cuneiform and 1st metatarsal, (*c*) the short muscles of the foot tying the two ends of the arch together, (*d*) the long flexor tendons which probably have a small supplementary action, (*e*) the peroneus longus acting as a sling below the lateral longitudinal arch.

The transverse arch

The cuboid, the lateral and intermediate cuneiforms and the bases of the metatarsals are wedge-shaped with the base of the wedge on the dorsal side. The highest point of the arch lies at the intermediate cuneiform and the base of the 2nd metatarsal. The ligamentous supports are the strong plantar and interosseous ligaments. The muscular support is the tendon of the peroneus longus passing across the sole of the foot from the lateral side of the cuboid to the base of the 1st metatarsal, thereby drawing the two sides of the arch together. The tibialis posterior and adductor hallucis also help to support this arch (Figs. 52 and 53).

The interdependence of bone, ligament and muscle in the maintenance of the arches is again emphasised. Probably the muscles could not maintain the arches without the other two for any length of time. It should also be noted that the height of the arches varies considerably from person to person without apparent impairment of function. In babies the bony arch is present from birth, but is frequently masked by fat in the sole of the foot. At birth inversion of the feet is so marked that the soles face each other; when walking begins eversion takes place with an accentuation of the medial side of the arch. The mobility of the arches is very variable.

CHAPTER 22

POSTURE AND WALKING

WEIGHT DISTRIBUTION IN THE ANATOMICAL POSITION

THE weight of the body is transmitted through the last lumbar vertebra to the sacrum. This tends to force the sacrum downwards and to rotate it so that the coccyx moves upwards and backwards. Downward movement of the upper part of the sacrum is counteracted by the extremely strong interosseous and dorsal ligaments of the sacro-iliac joints, assisted by the irregularities of the joint surfaces. Rotation of the sacrum is counteracted by the sacrospinous and sacrotuberous ligaments. The weight is transmitted through the irregular joint surfaces, along the thickened bodies of the ilia, to the acetabula. When sitting, the separation of the hip bones is counteracted by the anterior strut of the pubic bones.

The weight is distributed through the acetabula to the heads of both femurs, particularly to their superior surfaces. In each limb, the weight passes from the head along the neck to the body of the femur, producing cortical thickenings and patterns of trabeculae corresponding to the stresses and strains to which the bone is subjected. The weight passes down the bodies of the femur and tibia to the upper surface of the talus. From the talus it is distributed to the tuberosities of the calcaneus and the heads of the metatarsals in the ratio of 1 : 1. Swaying backwards and forwards at the ankle joints alters this ratio. The ratio of distribution within the metatarsals is 2 : 1 : 1 : 1 : 1, from the 1st to the 5th.

The body weight is balanced at the various joints with slight swaying movements and strictly speaking there is no such thing as a fixed posture. Individuals vary greatly in the way they perform these "standing movements." Thus the weight may be borne equally on both limbs or most of the weight may be carried on one limb and the other used as a prop, either crossed or uncrossed, to balance the body. In the anatomical position, the centre of gravity of the body is within the pelvis in front of the second segment of the sacrum. A vertical line through the centre is the line of weight (Fig. 60).

The line of weight of the trunk in the anatomical position passes slightly behind the plane of the centres of the hip joints. The trunk is prevented from falling further backwards by the anterior liga-

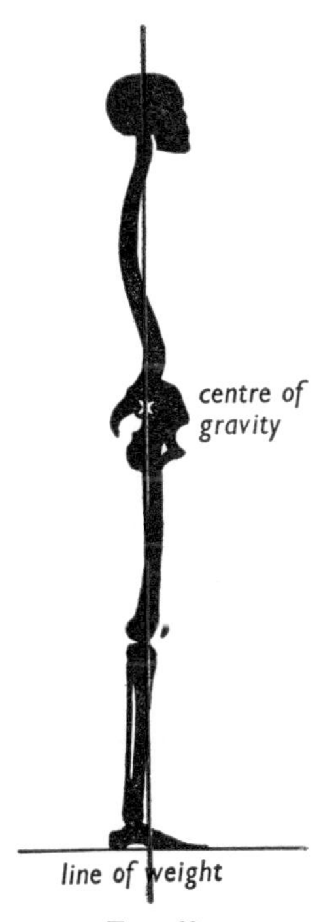

FIG. 60
In the anatomical position, the lateral projection of the line of weight passes through the external auditory meatus, cuts the vertebral column at different sites, passes just behind the centre of the hip joints, but in front of the centre of the knee joints, in front of the ankle joints, and midway between the calcaneal tuberosity and the heads of the metatarsals, the base on which we stand. The centre of gravity of the body is in the pelvis about the level of the 2nd sacral segment.

ments of the hip joint and possibly by contraction of the iliopsoas muscles.

At the knees the line of weight of that part of the body above the knee joints falls in front of the centre of the joints and the patella is movable indicating that the quadriceps femoris is relaxed. Falling of the body forwards is prevented by the posterior, lateral and medial collateral and cruciate ligaments of the knee. In other standing positions, the joint is not so completely extended and flexion at the knee causes the line to fall behind the transverse axis of the joints. Posture is maintained in this position by the contraction of the quadriceps femoris. This muscle is most important in the maintenance of stability at the knee joint since in many activities, such as standing up, sitting down or climbing, the line of weight falls behind the transverse axis of the joints.

At the ankles, the line of weight falls in front of the joints and therefore the weight tends to produce dorsiflexion. This is counteracted by the calf muscles, particularly the soleus.

WALKING

In walking each limb has a **stance** (supporting) **phase** and a **swing phase** (Fig. 61). The stance phase begins as the heel touches the ground (heel strike) (Fig. 61A) and one step by the right limb is completed when the right heel meets the ground again (Fig. 61E). At heel strike the thigh is flexed at the hip, the leg is extended at the knee and the foot is dorsiflexed at the ankle. During the first part of the stance phase there is extension at the hip associated with contraction of the hamstrings and gluteus maximus, slight flexion at the knee accompanied by contraction of the quadriceps femoris and plantar flexion of the foot (Fig. 61B). The weight of the body is transferred from the heel along the outer border of the foot to the heads of the metatarsals. In the later part of the stance phase (Fig. 61C), extension at the hip increases but without contraction of the extensor muscles. There is also extension at the knee without contraction of the quadriceps femoris until the end of the stance phase, when this muscle frequently contracts. At the same time the body becomes dorsiflexed over the foot. As this occurs the heel rises and the whole weight of the body is transferred to the heads of the metatarsals and then medially to the big toe. The raising of the heel is associated with plantar flexion of the body on the foot and the contraction of the calf muscles (soleus and gastrocnemius).

In the early part of the stance phase of the right limb, the limb is on the ground. As the left limb moves into its swing phase the whole weight of the body is on the right limb. The body tends to fall to the left and this is prevented by contraction of the right gluteus medius and minimus. They remain contracted during the whole of the swing phase of the left limb (Fig. 61 B and C).

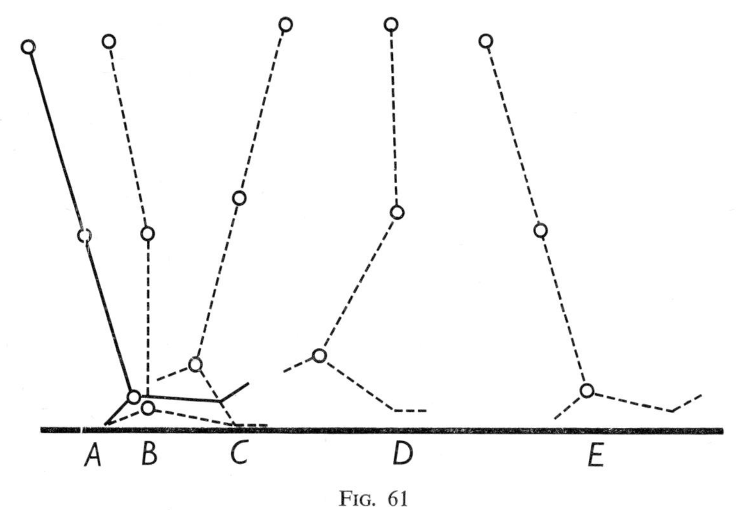

FIG. 61
The movements involved in walking. For description, see the text.

At the end of its stance phase, the right limb enters its swing phase (Fig. 61D) in which the thigh is flexed at the hip associated with contraction of the iliopsoas. This muscle, however, relaxes towards the end of flexion at the hip. During the swing phase there is flexion followed by extension at the knee. In many subjects both movements are unaccompanied by contraction of the appropriate muscles (hamstrings and quadriceps femoris) although some subjects show contraction of the quadriceps femoris just before heel strike (Fig. 61E). In the swing phase there is dorsiflexion of the foot at the ankle, often accompanied by contraction of the dorsiflexors only at the beginning and end of the phase (Fig. 61 C, D, E). During the swing phase the right ilium moves in front of the left ilium and, to keep the right foot parallel with the line of advance, the right limb is laterally rotated at the hip joint.

In walking, although one limb is alternately in the stance and swing phases, both limbs are twice on the ground at the same time for part of each step. Starting to walk from the anatomical position involves a forward tilting of the body. One limb flexes for the swing phase and the other extends for the second half of the stance phase.

It must be realised that there are no clear-cut static positions in the movements described above and that they merge imperceptibly into each other. Movements at different stages are produced by both muscle effort and body momentum. Extra muscle effort is required to overcome the initial inertia but little force is needed to maintain the movement. A smooth walking action depends on a series of pushes and checks so as to make the maximum use of gravity. The limbs act as flexible levers so arranged as to provide a rhythmic action by which the body moves forward. The body can be pushed uphill and lowered downhill as well as moved over uneven surfaces. Limitation of movement at any joint, as a result of injury or disease, alters the gait markedly.

CHAPTER 23

THE CUTANEOUS NERVE SUPPLY OF THE LOWER LIMB

THE cutaneous nerve supply is considered in relation firstly to the segmental distribution from the spinal cord (Fig. 62), and secondly to the peripheral nerve branches (Fig. 63). The rearrangement of nerve fibres in the lumbosacral plexus accounts for the representation in the peripheral nerves of the limbs of several segments of the spinal cord. The areas supplied by the cutaneous

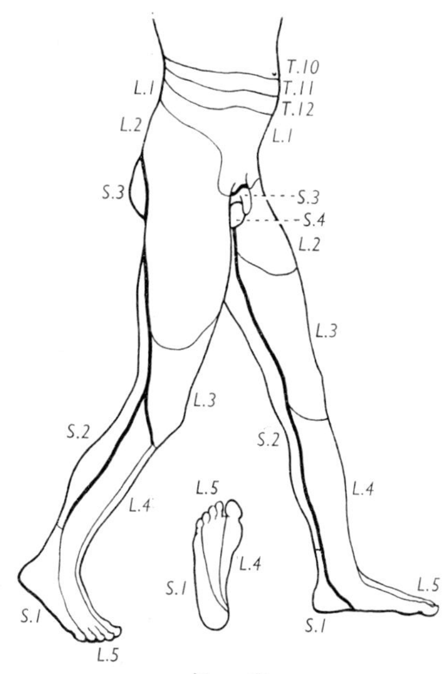

FIG. 62
The segmental cutaneous nerve supply of the lower limb. Thick black lines indicate the positions of the dorsal (on right leg) and ventral (on left leg) axial lines.

nerves have a considerable overlap so that section of any one nerve leads to only a limited loss of sensory function in the centre of the area supplied. The peripheral portions are able to function because of the supply from neighbouring nerves.

The segmental distribution can be regarded as starting with the 12th thoracic nerve about the level of the anterior superior iliac spine. The 1st lumbar nerve supplies the groin and the upper part of the buttock, the 2nd lumbar nerve the lateral and anterior aspects of the thigh and the lateral part of the buttock, and the 3rd lumbar nerve the medial side of the thigh and knee.

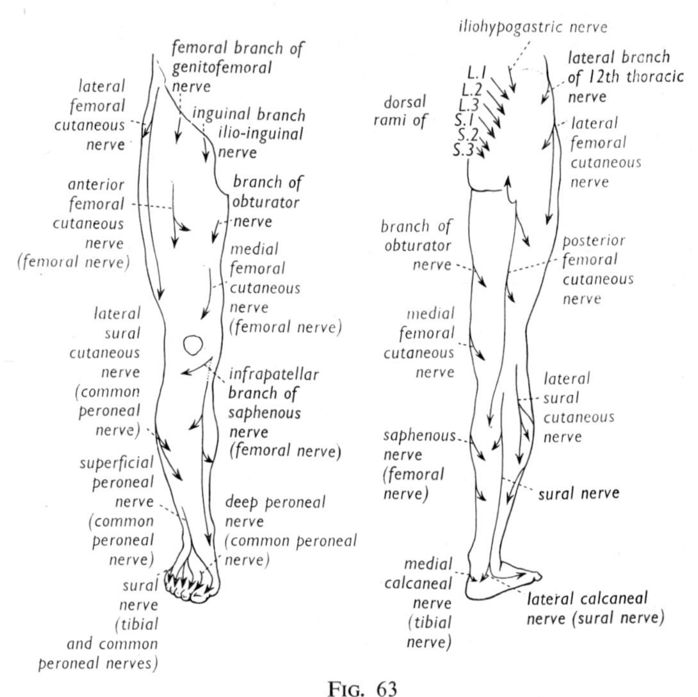

FIG. 63
The peripheral cutaneous nerve supply of the lower limb.

Below the knee the 4th lumbar nerve supplies the medial side of the calf, ankle and foot, and the 5th lumbar nerve the lateral side of the calf and part of the dorsal and plantar surfaces of the foot. The 1st sacral nerve supplies the lateral border of the foot and lower half of the back of the calf, the 2nd sacral nerve a strip along the back of the thigh almost to the gluteal fold, the 3rd sacral nerve the medial part of the buttock and the penis, and the 4th sacral nerve the scrotum and anus.

The larger cutaneous nerves are shown in Fig. 63.

CHAPTER 24

THE VENOUS AND LYMPH DRAINAGE OF THE LOWER LIMB

VEINS

THE limb has two systems of veins, superficial and deep, both of which have valves. The superficial veins lie outside the deep fascia and the deep ones are deep to it. For most of their course the latter lie between muscles. The two systems communicate by perforating (anastomotic) veins mainly in the regions of the thigh, knee and leg. They offer alternative routes for the upward movement of blood. The pressure of contracting muscles squeezes the blood up the deep veins from one intervalvular segment to the next. If all the blood cannot easily pass through the deep veins, the superficial veins take some of the flow and may even become overdistended. On the other hand, if no muscular contraction takes place the pumping action of the muscles on the deep veins is absent. The blood therefore tends to gravitate downwards and fill the deep venous system. Normally the blood passes via the perforating veins from the superficial to the deep veins. Incompetence of this mechanism is thought to be one of the causes of varicose veins.

The superficial veins (Fig. 64)

On the dorsum of the foot is a well-marked **dorsal venous arch.** This collects blood from the toes, from the dorsum of the foot and by perforating branches from the sole of the foot. From the medial and lateral sides of the arch arise the great and small saphenous veins respectively. The **great saphenous vein** runs upwards in front of the medial malleolus, crosses the lower third of the tibia, and ascends along the medial side of the leg, knee and thigh. It receives tributaries from the front and back of the thigh. As it passes through the saphenous opening to enter the femoral vein it receives a number of tributaries from the region of the groin.

The **small saphenous vein** arises from the lateral side of the dorsal venous arch, and passes upwards behind the lateral malleolus. It passes to the midline of the back of the calf, draining the

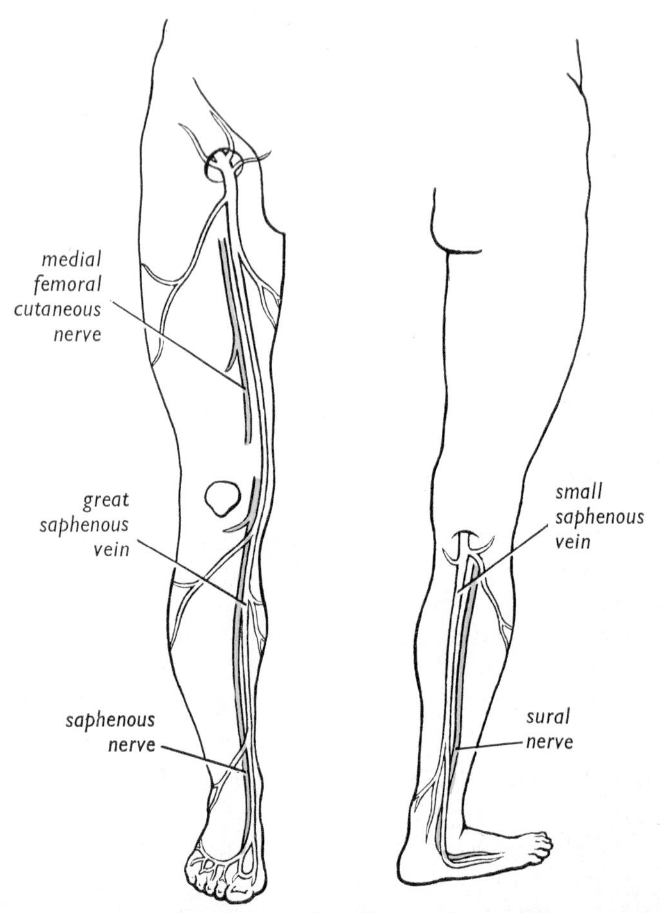

Fig. 64
The superficial venous drainage of the lower limb.

back and lateral sides of the leg, and communicating freely with the great saphenous vein. It runs upwards to the back of the knee. pierces the deep fascia and enters the popliteal vein.

The deep veins

These follow the main arteries, draining ultimately via the femoral vein into the external iliac vein. The veins accompanying the gluteal arteries drain into the internal iliac vein.

LYMPH DRAINAGE

The main groups of lymph nodes of the lower limb are arranged as follows:

1. The **upper superficial inguinal group** lies just distal and parallel to the inguinal ligament. This group drains very little of the limb, even though situated within it. Some of the superficial lymph vessels of the gluteal region drain to the lateral nodes of this group but the group as a whole drains the lower quadrant of the abdominal wall, and the perineum, penis, spongy urethra (the labia and the lower third of the vagina in the female) and the anal canal. The efferents of this group drain into groups 2 and 3.

2. The **lower superficial inguinal group** lies in front of the upper part of the femoral artery in the superficial fascia. It receives lymph from almost the whole of group 1 and from the vessels accompanying the great saphenous vein from the superficial parts of the leg other than the parts drained by group 4. The nodes drain into group 3 by vessels passing through the deep fascia at the saphenous opening.

3. The **deep inguinal group** lies around the femoral sheath deep to the deep fascia. The nodes receive efferent vessels from the lower superficial inguinal group and deep vessels from the whole limb. They drain into the external iliac group, some of the vessels passing through the femoral canal.

4. The **popliteal group** is a small group in the popliteal fossa. The nodes receive vessels from the lateral side of the calf and pass with the deep vessels to the deep inguinal group.

The superficial lymph vessels, starting from dorsal and plantar plexuses on the toes, drain into the lower superficial inguinal group except for a few on the lateral side which drain into the popliteal group. They tend to accompany the superficial veins.

The deep lymph vessels normally accompany the deeper blood vessels and pass to the deep inguinal group.

PRACTICAL CLASS 1

This is a guide to what should be known about individual bones and the articulated skeleton.

THE OSTEOLOGY OF THE PELVIS AND FEMUR

Requirements: articulated pelvis and femurs; separate hip bone, sacrum and femur.

1. **The hip bone**

Place the hip bone in the position it would occupy in the upright body and identify .

1. the anterior superior and anterior inferior iliac spines,
2. the iliac crest and its tubercle,
3. the posterior superior and posterior inferior iliac spines,
4. the gluteal surface of the ala of the ilium,
5. the greater and lesser sciatic notches,
6. the ischial spine and tuberosity,
7. the ischial ramus and inferior pubic ramus (the ischiopubic ramus),
8. the body of the pubis with the pubic crest and tubercle,
9. the symphysis pubis,
10. the superior pubic ramus with the pectineal surface,
11. the sacral area (articular and ligamentous) of the ilium,
12. the obturator foramen,
13. the acetabulum and acetabular notch.

2. **The sacrum**

Place the sacrum in the position it would occupy in the upright body and identify

1. the pelvic sacral foramina, the vertebral bodies medially, the lateral parts,
2. the promontory,
3. the dorsal sacral foramina,
4. the spinous tubercles on the median sacral crest,
5. the sacral cornua and hiatus.
6. the articular surfaces for (i) the ilium, (ii) the last intervertebral disc, (iii) the last lumbar vertebra, (iv) the coccyx.

3. The femur

Place the femur in the position it would occupy in the upright body.

(*a*) Comment on
1. the direction of the neck,
2. the neck-body angle,
3. the bowing of the body,
4. the obliquity of the body.

(*b*) On the upper end identify
1. the head with the pit for the ligament on the medial side,
2. the neck,
3. the greater and lesser trochanters,
4. the intertrochanteric line and crest,
5. the gluteal tuberosity.

(*c*) On the body identify
1. the linea aspera,
2. the medial and lateral supracondylar lines,
3. the adductor tubercle.

4. Muscular attachments

Define the approximate bony attachments of the following muscles:

(*a*) the flexors of the thigh,
1. the iliopsoas,
2. the pectineus,
3. the rectus femoris,
4. the sartorius.

(*b*) the extensors of the thigh,
1. the gluteus maximus,
2. the semimembranosus, semitendinosus and biceps femoris.

(*c*) the abductors of the thigh,
1. the gluteus medius and minimus,
2. the tensor fasciae latae.

(*d*) the adductors of the thigh,
1. the adductors longus, brevis and magnus,
2. the gracilis.

(*e*) the lateral rotators of the thigh,
1. the piriformis,
2. the obturator internus,
3. the quadratus femoris,
4. the obturator externus.

(*f*) the medial rotator of the thigh,
1. the gluteus minimus.

5. **Ligamentous attachments**

Define the bony attachments of the following ligaments:

(*a*) of the pelvis

1. the iliolumbar,
2. the ventral and dorsal sacro-iliac,
3. the sacrospinous and sacrotuberous,
4. the inguinal, lacunar and pectineal ligaments.

(*b*) of the hip joint

1. the capsule,
2. the iliofemoral,
3. the pubofemoral,
4. the ischiofemoral,
5. the acetabular labrum,
6. the transverse ligament of the acetabulum,
7. the ligament of the head of the femur.

PRACTICAL CLASS 2

THE ANATOMY OF THE HIP IN THE LIVING SUBJECT

Requirements: Skin pencils and tape measures, subjects.

1. Find the following bony landmarks:
   1. the anterior superior iliac spine,
   2. the posterior superior iliac spine,
   3. the highest points of the iliac crests,
   4. the ischial tuberosity,
   5. the upper edge of the greater trochanter on the outer aspect of the thigh,
   6. the upper edge of the symphysis pubis.
   7. the pubic tubercle.
2. With these landmarks as your guides, carry out the following exercises.

(*a*) Join the highest palpable points of the iliac crests (the supracristal plane) and note that this line lies between the spinous processes of the 3rd and 4th lumbar vertebrae.

(*b*) Examine the surface contours of the thigh, starting with the upper part in front. The junction of the thigh and the abdomen is marked by a shallow groove extending from the anterior superior iliac spine to the pubic tubercle, that is, along the line of the inguinal ligament which is attached to these two bony points. Below the middle third of the inguinal ligament is a shallow depression, which is more marked when the thigh is flexed and laterally rotated, and indicates the position of the femoral triangle. Ask the subject to adduct the thigh against resistance and show that the medial boundary of the triangle is formed by the prominence of the adductor longus. Ask the subject to flex and laterally rotate the thigh against resistance and show that the lateral boundary of the triangle is formed by the sartorius.

(*c*) On the medial side of the thigh show the adductor muscle mass by making the subject adduct the thigh against resistance. The adductor longus has been felt already. The muscle mass deep to this is mainly the adductor magnus. Trace its lowest fibres to the adductor tubercle.

(*d*) Measure the length of the limb. With the subject lying supine find the tip of the medial malleolus (the prominence on the medial side of the lower end of the tibia). Measure the distance from this point to (1) the upper edge of the greater trochanter, (2) the anterior superior iliac spine, (3) the umbilicus.

(i) Are these measurements the same on both sides?
(ii) Are these measurements altered by tilting of the pelvis?
(iii) How can one ascertain that the pelvis is not tilted?

(*f*) A shallow depression running down the outer surface of the thigh marks the iliotibial tract. At its upper end in front of the depression lies the tensor fasciae latae and behind it the gluteus maximus. Ask the subject to stand on one limb and palpate the gluteal muscles while their contours are observed.

(*g*) With the subject standing upright, mark the two anterior superior iliac spines. Is there any change in their relative positions when he stands on one lower limb?

(*h*) With the subject standing upright examine the gluteal fold. This lies horizontally and tends to disappear when the thigh is flexed. Confirm that it is not formed by the lower edge of the gluteus maximus.

(*i*) Determine the extent of the following movements at the hip joint and name the prime movers of each. What is the effect of extension at the knee on flexion at the hip?

1. abduction,
2. adduction,
3. flexion,
4. extension,
5. medial rotation,
6. lateral rotation.

3. The surface markings of the following structures should be noted.

(*a*) The **femoral artery.** The femoral point is midway between the symphysis pubis and the anterior superior iliac spine. The femoral artery can be palpated from this point along the upper two-thirds of a line drawn to the adductor tubercle with the thigh slightly flexed and laterally rotated.

(*b*) The **femoral vein** lies immediately medial to the artery.

(*c*) The **femoral canal** lies medial to the vein and just distal to the inguinal ligament.

(*d*) The **femoral nerve** lies just lateral to the artery.

(*e*) The **saphenous opening** lies superficial to the femoral vein, 4 cm distal and lateral to the pubic tubercle. In some subjects a dilated great saphenous vein may be seen passing up the medial side of the thigh towards the saphenous opening.

(*f*) The **sciatic nerve** enters the thigh slightly medial to the mid-point of a line joining the greater trochanter and the ischial tuberosity. In a thin subject the nerve can be felt as a rounded cord against the ischium. Above this point it curves medially, and below this point it runs down the centre of the back of the thigh.

PRACTICAL CLASS 3

THE KNEE

Requirements: Skeleton, femur, patella, tibia, fibula and skin pencils.

THE OSTEOLOGY OF THE REGION OF THE KNEE

1. The femur

Place the femur in the position it would occupy in the upright body. On the lower end

(*a*) identify

1. the medial condyle and epicondyle,
2. the lateral condyle and epicondyle,
3. the intercondylar fossa,
4. the medial and lateral supracondylar lines.

(*b*) outline the areas on the articular surface for the patella and the tibial condyles.

(*c*) outline the capsular attachment.

2. The patella

Place the patella as it would be in the upright body and identify

1. the articular and non-articular surfaces,
2. the curved edge for the attachment of the tendon of the quadriceps femoris,
3. the pointed edge (apex) for the attachment of the patellar ligament,
4. the larger lateral and smaller medial femoral articular surfaces.

3. The tibia

Place the tibia in the position it would occupy in the upright body. On the upper end identify

1. the medial and lateral condyles,
2. the tibial tuberosity at the proximal end of the anterior border,
3. the anterior, medial and interosseous borders,
4. the articular facet for the upper end of the fibula,
5. the articular surfaces for the lower end of the femur,
6. the intercondylar eminence,
7. the soleal line on the posterior surface,
8. the groove on the back of the medial condyle.

4. The fibula

Place the fibula in the position it would occupy in the upright body and identify

(*a*) the lower end with the lateral malleolar fossa, which lies posterior and medial, and the talar and tibial articular areas,

(*b*) the upper end (head) with its apex and articular facet for the tibia. (Note its oblique plane.)

5. On the bones outline the areas of attachment of the following muscles:
   1. the quadriceps femoris,
   2. the posterior femoral muscles (hamstrings),
   3. the gastrocnemius,
   4. the popliteus,
   5. the sartorius and gracilis.

6. Where are the following structures attached?
   1. the tibial collateral ligament of the knee,
   2. the fibular collateral ligament of the knee,
   3. the patellar ligament,
   4. the anterior and posterior cruciate ligaments,
   5. the medial meniscus,
   6. the lateral meniscus.

THE ANATOMY OF THE KNEE JOINT IN THE LIVING SUBJECT

(*a*) Find on the subject: the adductor tubercle, the medial and lateral condyles and epicondyles of the femur; the patella; the tuberosity and condyles of the tibia; and the head and neck of the fibula.

(*b*) Notice the alteration in the position of the patella as the knee is flexed. The lateral edge of the patella can be felt to tilt forwards and the medial condyle can be felt medial to the patella. The groove corresponding to the joint between the condyles of the femur and tibia can also be felt laterally in flexion. Extend the knee and feel the mass of the quadriceps femoris on the front of the thigh. Follow it down the lateral side and note the flattening of the muscle. On the other side, there is a bulge just above and medial to the patella, formed by the lowermost fibres of the vastus medialis. Follow the patellar ligament from the apex of the patella to the tibial tuberosity.

(*c*) Behind the knee examine the popliteal fossa. Note the diamond-shaped hollow when the knee is partially flexed. Above, feel the tendon of the biceps femoris on the lateral side and the round tendon of the semitendinosus lying on the semimembranosus on the medial side. The tendon of the adductor magnus may be felt in the hollow between the semimembranosus and the vastus medialis. The medial and lateral heads of the gastrocnemius, the inferior boundaries of the space, are not as clearly defined as the hamstring tendons above the joint.

(*d*) Feel the pulsation of the popliteal artery deep in the middle of the space. It is most easily felt with the knee flexed.

(*e*) Follow the tendon of biceps femoris downwards to the head of the fibula and palpate the common peroneal nerve as a horizontal cord distal to the tendon as the nerve passes forwards round the neck of the fibula. Anterior to the biceps, the iliotibial tract can be felt.

(*f*) Extension at the knee takes place when the leg is brought forwards on the thigh or when rising from the sitting position. With the leg fully extended at the knee the greater trochanter, the lateral epicondyle and the lateral malleolus are in the same coronal plane.

(*g*) Flexion occurs when the calf of the leg is brought towards the posterior surface of the thigh. Demonstrate again that flexion at the hip is greater with the knee flexed than with the knee extended.

PRACTICAL CLASS 4

THE LEG AND ANKLE

Requirements: Skeleton, tibia, fibula, foot bones and skin pencils.

THE OSTEOLOGY OF THE LEG AND ANKLE JOINT

1. The tibia

Place the tibia in the position it would occupy in the upright body. On the lower end identify

1. the subcutaneous surface,
2. the medial malleolus,
3. the fibular notch,
4. the articular surfaces for the talus and the fibula,
5. the groove for the tibialis posterior tendon.

2. **The fibula**

Place the fibula in the position it would occupy in the upright body. On the lower end identify

1. the lateral malleolus,
2. the lateral malleolar fossa,
3. the interosseous border,
4. the articular surfaces for the tibia and talus,
5. the groove for the peroneal tendons.

3. Show the approximate attachments of the following on the tibia and / or the fibula

1. the patellar ligament,
2. the biceps femoris, gracilis, semimembranosus, semitendinosus and sartorius,
3. the extensor hallucis longus and extensor digitorum longus, peroneus tertius and tibialis anterior,
4. the peroneus longus and brevis,
5. the popliteus and soleus,
6. the flexor hallucis longus, flexor digitorum longus and tibialis posterior,
7. the interosseous membrane and ligament.

4. **The talus**

Place the talus in the position it would occupy in the upright body and identify

1. the head, neck and body,
2. the articular surfaces for the lateral and medial malleoli and the tibia (trochlea),
3. the articular surfaces for the calcaneus, the navicular bone and the plantar calcaneonavicular ligament,
4. the sulcus tali,
5. the lateral and medial tubercles,
6. the groove for the flexor hallucis longus.

5. **The calcaneus**

Place the calcaneus in the position it would occupy in the upright body and identify

1. the upper surface with the sulcus calcanei and the articular facets for the talus,

2. the lower surface with its tuberosity,
3. the posterior surface with the area for the attachment of the tendo calcaneus,
4. the anterior surface with the articular surface for the cuboid,
5. the medial surface with the sustentaculum tali,

6. Articulate the calcaneus with the talus and show the sinus tarsi.

THE ANATOMY OF THE LEG AND ANKLE IN THE LIVING SUBJECT

1. Palpate the tuberosity of the tibia and note that it is about 2 cm distal to the level of the tibial condyles. Follow the anterior border of the tibia from the tuberosity as far as the anterior border of the medial malleolus. Both malleoli are subcutaneous; the lateral malleolus is longer, larger and more prominent than the medial malleolus and its tip is 2 cm distal to that of the medial malleolus. Proximal to the medial malleolus the whole of the smooth, medial, subcutaneous surface of the tibia can be felt.

   The fibula cannot be palpated except at its upper and lower ends.
2. On the subject palpate and mark
   1. the tendons of the tibialis anterior, extensor hallucis longus, extensor digitorum longus,
   2. the position of the anterior tibial artery in front of the ankle joint,
   3. the tendons of the peroneus longus and brevis,
   4. the tendo calcaneus,
   5. the posterior tibial artery,
   6. the saphenous veins with the subject standing upright. Mark as much of their course as you can see.

3. Palpate the muscle bellies in the middle of the calf and ask the subject to stand on his toes. Note the change in the muscle.

4. The movements at the ankle joint are dorsiflexion and plantar flexion. In dorsiflexion the dorsum of the foot is brought towards the anterior surface of the leg. Estimate its extent. Plantar flexion is movement in the opposite direction.

5. Make the subject plantar flex the foot with the knee extended and note the contraction of the superficially placed gastrocnemius. Now make him bend the knee and again plantar flex the foot. The gastrocnemius does not contract but the soleus, palpable in front of each side of the gastrocnemius, contracts. On the lateral side of the leg, the peroneus longus and brevis contract when the foot is everted against resistance.
6. On each side of the tendo calcaneus is a well-marked hollow. On the medial side between the tendo calcaneus and the medial malleolus the tendons of the flexor hallucis longus and flexor digitorum longus can be felt on plantar flexing the toes against resistance, and immediately posterior to the malleolus the tibialis posterior is prominent and easily felt on inversion and plantar flexion. On the anterior aspect of the ankle joint, the tendon of the tibialis anterior is on the medial side and is palpable on dorsiflexion. The extensor hallucis longus tendon can be felt immediately lateral to this on dorsiflexion of the great toe, and extensor digitorum longus tendon can be felt when the toes are dorsiflexed against resistance. The peroneus tertius may be felt when the foot is everted against resistance.

PRACTICAL CLASS 5

THE FOOT

Requirements: Skeleton of the leg and foot, skin pencils.

THE OSTEOLOGY OF THE FOOT

1. **The navicular**

   On the navicular show
   1. the articular facets for the talus and the cuneiforms,
   2. the tuberosity on the medial side.

2. **The cuneiforms**

   On the cuneiforms show
   1. the large medial bone with a broad plantar surface,
   2. the smaller middle and lateral bones with broad dorsal surfaces,

3. the articular facets for the navicular and the three medial metatarsals.

3. **The cuboid**

On the cuboid show

1. the articular facets for the calcaneus, the lateral cuneiform and the lateral two metatarsals,
2. the groove on the plantar surface.

4. **The metatarsals**

On the metatarsals show

1. the short, strong first bone,
2. the long, thin second bone,
3. the tuberosity on the fifth bone.

5. **The phalanges**

On the phalanges note that

1. the first toe is the largest,
2. the second toe is often the longest,
3. the terminal phalanges are very small.

6. Show the approximate attachments of the following muscles
   1. the tibialis anterior, extensor hallucis longus, extensor digitorum longus and extensor digitorum brevis,
   2. the tibialis posterior, flexor hallucis longus, flexor digitorum longus and flexor accessorius,
   3. the peroneus longus and peroneus brevis.

7. Define the attachments of the following
   1. the plantar calcaneonavicular (spring) ligament,
   2. the plantar calcaneocuboid ligament,
   3. the long plantar ligament,
   4. the plantar aponeurosis.

THE ANATOMY OF THE FOOT IN THE LIVING SUBJECT

1. Examine the movements of inversion and eversion. These occur in the foot and not at the ankle. What muscles produce these movements?

2. Plantar flexion and dorsiflexion of the toes are easily demonstrated, but abduction and adduction are frequently limited.
3. Identify the malleoli, the head of the talus, the sustentaculum tali, the navicular tuberosity, the tuberosity of the fifth metatarsal and the peroneal trochlea.
4. Observe the arches of the foot with and without weight-bearing.
5. Grip the heel with one hand and palpate the head and neck of the talus with the other hand. Demonstrate that inversion and eversion take place mainly at the talocalcaneal joints. Does it matter whether the foot is plantar or dorsiflexed?
6. With the subject walking slowly, note the movements at the different joints and suggest the various groups of muscles which may produce these movements.

INDEX

THE UPPER AND LOWER LIMBS

Printed by The Central Press (Aberdeen) Ltd.